AF358647

Machining Dynamics and Parameters
Process Optimization

Machining Dynamics and Parameters Process Optimization

Editors

Gorka Urbicain
Daniel Olvera Trejo

MDPI • Basel • Beijing • Wuhan • Barcelona • Belgrade • Manchester • Tokyo • Cluj • Tianjin

Editors
Gorka Urbicain
University of the Basque Country (UPV/EHU)
Spain

Daniel Olvera Trejo
Tecnologico de Monterrey
Mexico

Editorial Office
MDPI
St. Alban-Anlage 66
4052 Basel, Switzerland

This is a reprint of articles from the Special Issue published online in the open access journal *Applied Sciences* (ISSN 2076-3417) (available at: https://www.mdpi.com/journal/applsci/special_issues/Machining_Dynamics).

For citation purposes, cite each article independently as indicated on the article page online and as indicated below:

LastName, A.A.; LastName, B.B.; LastName, C.C. Article Title. *Journal Name* **Year**, *Volume Number*, Page Range.

ISBN 978-3-0365-0294-6 (Hbk)
ISBN 978-3-0365-0295-3 (PDF)

Contents

About the Editors

Gorka Urbicain, Assistant Professor at University of the Basque Country. He studied Mechanical Engineering at the Faculty of Engineering of Bilbao. During his Ph.D. thesis, he worked in the field of mathematical modeling of stability in turning systems in the Department of Mechanical Engineering at the University of the Basque Country. He is the author of more than 30 contributions in JCR journals (Q1 + Q2) and international conferences and actively collaborates in international/national research projects in the field of manufacturing.

Daniel Olvera Trejo, Assistant Professor at Tecnológico de Monterrey. He is a researcher in the School of Engineering and Sciences and teaches at the Department of Mechanics Engineering and Advanced Materials. His research interests focus mainly on the implementation of optimization algorithms for vibration avoidance in the machining process. He has collaborated on R&D projects related to aerospace components machining, tooling design, and machine tools vibration characterization. He has authored over 30 scientific papers with more than 300 citations.

 *applied*
sciences

Editorial

Special Issue on "Machining Dynamics and Parameters Process Optimization"

Gorka Urbikain [1,*] **and Daniel Olvera-Trejo** [2,*]

[1]   Department of Mechanical Engineering, University of the Basque Country, Alameda de Urquijo s/n, 48013 Bilbao, Bizkaia, Spain
[2]   Tecnológico de Monterrey, Escuela de Ingeniería y Ciencias, Av. Eugenio Garza Sada 2501, Monterrey, Nuevo León 64849, Mexico
*   Correspondence: gorka.urbikain@ehu.eus (G.U.); daniel.olvera.trejo@tec.mx (D.O.-T.)

Received: 8 December 2020; Accepted: 10 December 2020; Published: 14 December 2020

1. Introduction and Scope

In 1907, F.W Taylor—the father of production engineering—exposed the fundamentals of modern machining and defined chatter as the most obscure and delicate of all problems facing the machinist. However, a century later, chatter is still a hot topic at the forefront of machining problems. New casted materials with higher properties lead to challenging designs with even thinner walls and floors. Despite nowadays new processes emerge, parts still need to be finished with metal removal processes as ever if aiming for good quality and close tolerances. Specifically, this book, *Machining Dynamics and Parameters Process Optimization*, includes a series of ten studies with some new advances regarding machining dynamics in metal cutting. The topics include the dynamic characterization of machine tools, experimental dampening techniques, or optimization algorithms combined with signal monitoring.

2. Contributions

All the contributions were divided into three different sub-topics.

2.1. Dynamic Characterization of Machine Tools and Stability Analysis

Unstable vibrations in machining processes arise due to the regenerative effect of the uncut chip thickness. Stability analysis enables the prevention of these undesirable vibrations, improving tool life, surface roughness and, in general, productivity. For instance, Alvarez et al. [1] present an innovative method for calculating stability maps in cylindrical and centerless infeed grinding process. The method is based on the application of the Floquet theorem by repeated time integrations. Additionally, Sosa et al. [2], discussed an updated method of the enhanced multistage homotopy perturbation method for the solution of delay differential equations (DDE) with multiple delays. The authors demonstrated better convergence rates and less computation time for the calculation of the stability lobes for multivariable cutters in the milling process. This method is also a very promising tool that could be extended to other machining processes.

The dynamic characterization of machine tools also plays an important role in understanding and increasing performance. This task demands accurate dynamic models and sophisticated approaches to characterize new machine tools. In the industrial scenario, the machining performance with robots is highly dependent on their dynamics; Sun et al. [3] introduced a method to predict natural frequencies of industrial robots that brings new knowledge for improving a 6R Robot's dynamics.

2.2. Devices and Experimental Techniques

The productivity during the machining of thin-floor and thin-walled components is limited due to unstable vibrations that are likely to occur when increasing depth of cut. Puma-Araujo et al. [4]

designed a custom-built semi-active magnetorheological damper device that increases the damping ratio and modal stiffness of a thin-floor workpiece, significantly expanding the stable areas when milling with a bull-nose cutter. A path is given for extending this practice towards industrial use.

Another source of vibration in five-axis machining is the discontinuities of the linear-segment toolpath inducing fluctuation in the programmed feed rate. Gao et al. [5] proposed a double B-spline curve-fitting and synchronization-integrated feed rate scheduling method. They experimentally concluded that this method generates a smooth toolpath with an allowable fitting error. The optimal setting of the computer numerical control (CNC) plays an important role in the machining performance; specifically, Yu et al. [6] developed a practical methodology to tune the CNC parameters effectively for easy implementation in a commercial CNC controller. They studied three turning machining modes: high-speed, high-accuracy and high-surface quality in terms of dynamics errors. Similarly, the accuracy determined by the CNC in a flute-grind of an end-mill, has an impact on the performance of the cuter. Fang et al. [7] proposed a new projection model intending to optimize the wheel's location and orientation for the flute-grinding to achieve higher accuracy and efficiency.

2.3. Monitoring Systems and Machining Optimization

Contemporary additive manufacturing (AM) has demonstrated the potential to manufacture lightweight, complex parts of which subtracting manufacturing was incapable. However, most of these parts still require a machining stage to achieve the required dimensions and surface roughness. Accordingly, Grossi et al. [8], presented an approach to predict the dynamics of a thin-walled component produced using wire–arc–additive manufacturing. The knowledge of the workpiece dynamics evolution throughout the machining process is necessary for cutting parameter optimization.

Furthermore, innovative machining monitoring systems based on artificial intelligence (AI) can minimize machining errors and yield high productivity by predicting and avoiding adverse conditions that cause a lack of dimensional accuracy and tool breakage. Mamledesai et al. [9] proposed a tool condition monitoring framework that included different quality requirements and monitors if the tool produces conforming parts. They used computer vision, a convolution neural network (CNN), and transfer learning (TL) to detect changes in quality indicators. However, because Industry 4.0 is becoming a reality in all the manufacturing sectors, deep learning (DL) for big data has emerged as a key tool. Zhang et al. [10] presented a tool-wear monitoring method for complex part milling based on deep learning. The features were pre-selected based on the cutting force model and wavelet packet decomposition. The pre-selected cutting forces, vibration and condition features were introduced into a deep autoencoder for dimension reduction, and then a deep multi-layer perceptron was developed to estimate tool wear.

Funding: This research received no external funding.

Acknowledgments: The guest editors would like to thank the authors, the reviewers, and the editorial team of *Applied Sciences* for their valuable contributions to this special issue.

Conflicts of Interest: The authors declare no conflict of interest.

References

1. Alvarez, J.; Zatarain, M.; Barrenetxea, D.; Marquinez, J.I.; Izquierdo, B. Implicit Subspace Iteration to Improve the Stability Analysis in Grinding Processes. *Appl. Sci.* **2020**, *10*, 8203. [CrossRef]
2. Sosa, J.L.; Olvera-Trejo, D.; Urbikain, G.; Martinez-Romero, O.; Elías-Zúñiga, A.; Lacalle, L.N.L. Uncharted Stable Peninsula for Multivariable Milling Tools by High-Order Homotopy Perturbation Method. *Appl. Sci.* **2020**, *10*, 7869. [CrossRef]
3. Sun, J.; Zhang, W.; Dong, X. Natural Frequency Prediction Method for 6R Machining Industrial Robot. *Appl. Sci.* **2020**, *10*, 8138. [CrossRef]

4.   Puma-Araujo, S.D.; Olvera-Trejo, D.; Martínez-Romero, O.; Urbikain, G.; Elías-Zúñiga, A.; López de Lacalle, L.N. Semi-Active Magnetorheological Damper Device for Chatter Mitigation during Milling of Thin-Floor Components. *Appl. Sci.* **2020**, *10*, 5313. [CrossRef]

5.   Gao, X.; Zhang, S.; Qiu, L.; Liu, X.; Wang, Z.; Wang, Y. Double B-Spline Curve-Fitting and Synchronization-Integrated Feedrate Scheduling Method for Five-Axis Linear-Segment Toolpath. *Appl. Sci.* **2020**, *10*, 3158. [CrossRef]

6.   Yu, B.-F.; Chen, J.-S. Development of an Analyzing and Tuning Methodology for the CNC Parameters Based on Machining Performance. *Appl. Sci.* **2020**, *10*, 2702. [CrossRef]

7.   Fang, Y.; Wang, L.; Yang, J.; Li, J. An Accurate and Efficient Approach to Calculating the Wheel Location and Orientation for CNC Flute-Grinding. *Appl. Sci.* **2020**, *10*, 4223. [CrossRef]

8.   Grossi, N.; Scippa, A.; Venturini, G.; Campatelli, G. Process Parameters Optimization of Thin-Wall Machining for Wire Arc Additive Manufactured Parts. *Appl. Sci.* **2020**, *10*, 7575. [CrossRef]

9.   Mamledesai, H.; Soriano, M.A.; Ahmad, R. A Qualitative Tool Condition Monitoring Framework Using Convolution Neural Network and Transfer Learning. *Appl. Sci.* **2020**, *10*, 7298. [CrossRef]

10.   Zhang, X.; Han, C.; Luo, M.; Zhang, D. Tool Wear Monitoring for Complex Part Milling Based on Deep Learning. *Appl. Sci.* **2020**, *10*, 6916. [CrossRef]

Publisher's Note: MDPI stays neutral with regard to jurisdictional claims in published maps and institutional affiliations.

Article

Implicit Subspace Iteration to Improve the Stability Analysis in Grinding Processes

Jorge Alvarez [1,*], Mikel Zatarain [1], David Barrenetxea [1], Jose Ignacio Marquinez [1] and Borja Izquierdo [2]

[1] IDEKO, Basque Research and Technology Alliance, Pol. Industrial Arriaga 2, 20870 Elgoibar, Spain; mmzatarain@gmail.com (M.Z.); dbarrenetxea@ideko.es (D.B.); jimarquinez@ideko.es (J.I.M.)

[2] Department of Mechanical Engineering, University of the Basque Country UPV/EHU, Plaza Torres Quevedo 1, 48013 Bilbao, Spain; borja.izquierdo@ehu.eus

* Correspondence: jalvarez@ideko.es

Received: 9 October 2020; Accepted: 18 November 2020; Published: 19 November 2020

Featured Application: The calculation time reduction of the new methodology for obtaining the stability maps in grinding processes contributes to the industrialization of chatter avoidance technologies.

Abstract: An alternative method is devised for calculating dynamic stability maps in cylindrical and centerless infeed grinding processes. The method is based on the application of the Floquet theorem by repeated time integrations. Without the need of building the transition matrix, this is the most efficient calculation in terms of computation effort compared to previously presented time-domain stability analysis methods (semi-discretization or time-domain simulations). In the analyzed cases, subspace iteration has been up to 130 times faster. One of the advantages of these time-domain methods to the detriment of frequency domain ones is that they can analyze the stability of regenerative chatter with the application of variable workpiece speed, a well-known technique to avoid chatter vibrations in grinding processes so the optimal combination of amplitude and frequency can be selected. Subspace iteration methods also deal with this analysis, providing an efficient solution between 27 and 47 times faster than the abovementioned methods. Validation of this method has been carried out by comparing its accuracy with previous published methods such as semi-discretization, frequency and time-domain simulations, obtaining good correlation in the results of the dynamic stability maps and the instability reduction ratio maps due to the application of variable speed.

Keywords: stability; machining; chatter; grinding

---

1. Introduction

Grinding is a finishing process characterized by its ability to achieve workpieces with smooth surfaces and fine tolerances. One of the most important limitations in grinding processes is the self-excited vibration or chatter. The appearance of this regenerative effect leads to lower form accuracy and worse surface finish of the ground parts, loss of productivity, grinding machine damage or even wheel breakage.

Chatter in grinding occurs when the grinding forces change during the process due to the variations of the chip thickness in the presence of vibrations, then the corresponding delayed responses to the forces increase the undulations over the workpiece surface. The generative effect can appear on both the workpiece and the grinding wheel, adding complexity to the dynamic behavior. The undulations generated on the workpiece surface grow quite rapidly, whereas those generated in the grinding wheel surface grow more gradually [1]. Besides, when the vibration amplitude builds up to a certain limit, those undulations on the grinding wheel surface are removed through the dressing process. For these reasons, this paper is only focused on the workpiece regenerative chatter.

Different approaches have been developed to analyze this phenomenon and devise practical methods for suppressing these vibrations [2]. The first discussion of regenerative stability of a grinding process was done by Hahn using the Nyquist criteria [3]. Then, Snoeys and Brown [4] performed a similar analysis by means of a doubly regenerative closed-loop block diagram and obtained the characteristic equation of the process and a stability criterion. Thompson proposed an alternative approach obtaining solutions of double regenerative chatter for the stability boundary [5] and chatter growth [6]. Later, different time-domain simulation approaches were applied to grinding stability analysis and regenerative chatter prediction [7,8], considering also non-linear effects [9,10].

Alvarez et al. applied the semi-discretization method proposed by Insperger et al. for milling [11] and turning [12]—later used for selection of variable pitch for chatter suppression in face milling [13] to infeed cylindrical grinding process [14] and traverse cylindrical grinding process [15]—demonstrating the potential of this approach to analyze the application of variable speed to chatter avoidance with an efficient computing effort comparing to previous time-domain simulations. This periodically variable speed is one of the most extended techniques for chatter suppression in machining processes in general [12,16,17], and in grinding processes in particular. Inasaki [18] was the first to simulate the workpiece sinusoidal speed variation effect, concluding that chatter can be avoided with short periods and large amplitudes of variation. Barrenetxea et al. [19] and Alvarez et al. [20] demonstrated the application of this technique for infeed and through-feed centerless grinding processes, respectively, by devising a time-domain dynamic approach for generating stability maps whose axes were the amplitude and frequency of the sinusoidal variation. The analysis concluded that the sinusoidal shape is the optimal speed variation signal. Those maps were validated experimentally, but the most important limitation of that method was the very long time required for obtaining a complete stability map due to the large amount of simulations needed.

Therefore, the goal of this paper is to obtain a method for computation of the instability lobes in the time domain with much lower computing effort compared to traditional ones. The method is based on the implicit subspace iteration presented by Zatarain et al. [21,22] for milling process analysis. The improvement comparing with the semi-discretization method is achieved by removing the need for calculating the complete transition matrix whose eigenvalues give the stability degree of parameter combinations.

This method is applied both in cylindrical and centerless infeed grinding processes and validated by comparing the results with previously published ones by Alvarez et al. [14] and Barrenetxea et al. [19,23], respectively, analyzing the accuracy and the calculation time reduction. New formulation has been included in the method due to the geometrical instabilities that appear during the centerless grinding processes.

2. Materials and Methods

The stability criterion for discretized time domain simulations is based on the Floquet theorem [24], as shown by Insperger et al. [11] for milling process and Alvarez et al. for grinding process [14]. This theorem states that when the coefficient matrices are periodic, the solutions of the homogeneous time periodic delay Equation:

$$\dot{\mathbf{x}}(t) = \mathbf{A}(t)\mathbf{x}(t) + \mathbf{B}(t)\mathbf{x}(t - \tau) \tag{1}$$

have the general form

$$\mathbf{x}(t) = e^{\lambda t}\mathbf{p}(t) \tag{2}$$

where $\mathbf{p}(t)$ is also periodic with principal time period T and being $\mathbf{A}(t) = \mathbf{A}(t+T)$ and $\mathbf{B}(t) = \mathbf{B}(t+T)$. Then, the solutions are exponentially increasing or decreasing periodical functions and the decay is described by the characteristic exponent λ.

After discretization, the solution at a certain moment and the solution one period later are related by the transition matrix **Z**:

$$\mathbf{q}_{i+m} = \mathbf{Z}\mathbf{q}_i \tag{3}$$

being $\mathbf{q}_i$ the multidimensional state at time t_i and $m = \Delta t/T$. The eigenvectors of the transition matrix **Z** are the discretized counterpart of the principal solutions of (1) and the eigenvalues μ of **Z**, that are called characteristic multipliers, are related to the characteristic exponents [25]

$$\lambda = \frac{1}{T}\left(\ln|\mu| + i(\arg(\mu) + 2k\pi)\right) \tag{4}$$

where k corresponds to the harmonics involved.

The stability condition is that there are no eigenvalues with a magnitude larger than 1. Therefore, the interesting eigenvalues are those of highest magnitude, as they will characterize the stability of the process.

The formation of the transition matrix **Z** is usually a costly procedure. Zatarain et al. proposed a method based on implicit subspace iteration with the aim of avoiding the calculation of the transition matrix as well as the resolution of the corresponding eigenvalue problem [21]. Therefore, the computational cost can be reduced considerably.

Power iteration is the simplest way to obtain the highest magnitude eigenvalue and associated eigenvector of a matrix but is a relatively slow method. Subspace iteration consists of performing the power iteration of some vectors $\mathbf{S}_0$ representing a guess of the dominant eigenvectors.

$$\mathbf{S}_0 = [\mathbf{s}_{0,1}, \mathbf{s}_{0,2}, \mathbf{s}_{0,3}, \ldots, \mathbf{s}_{0,n}] \tag{5}$$

Initially there is no way to consider that the vectors in **S** are the eigenvectors corresponding to the largest magnitude eigenvalues, nor that they contain a large projection on these eigenvectors. Nevertheless, after each time period, the projection of all the vectors on the dominant eigenvectors will increase in proportion to their corresponding eigenvalues.

The use of a group of vectors (subspace **S**) instead of a single vector and the orthogonalization of these vectors with respect to the matrix **S** are the basis of the subspace iteration method and give rise to a reduction in the calculation time. Therefore, the procedure will consist of multiplying the subspace by the transition matrix Z once again and proceeding to the orthogonalization of the subspace after each or after several multiplications. The implicit subspace iteration method [21] follows that methodology but without calculating the transition matrix.

The multiplication of the subspace **S** by transition matrix **Z** is the first step and can be substituted by the time integration of the vectors in **S** by a fast procedure

$$\mathbf{V}_i = [\mathbf{v}_{i,1}, \mathbf{v}_{i,2}, \mathbf{v}_{i,3}, \ldots, \mathbf{v}_{i,n}] = \mathbf{Z}\mathbf{S}_i \tag{6}$$

Now it is necessary to proceed to the orthogonalization of the vectors in **S**. The proposed method performs the orthogonalization, calculating an **H** matrix as

$$\mathbf{H}_i = \left(\mathbf{S}_i^T\mathbf{S}_i\right)^{-1}\mathbf{S}_i^T\mathbf{V}_i \tag{7}$$

and then performing an eigenvalue decomposition on it as

$$\mathbf{H}_i = \mathbf{G}_i\lambda_i\mathbf{G}_i^{-1} \tag{8}$$

Finally, the calculation of the new subspace is obtained as

$$\mathbf{S}_{i+1} = \mathbf{G}_i\mathbf{V}_i \tag{9}$$

After enough iterations, the subspace **S** will converge to the dominant eigenvectors, and the eigenvalues of λ will be the dominant eigenvalues.

With this method, it is possible not to form the transition matrix **Z**, but just to perform the time integration in one period (corresponding to a workpiece revolution) and obtain the new set of vectors **V**. The transition matrix is not required either to perform the orthogonalization of the subspace vectors.

The method of semi-discretization for calculating the eigenvalues requires calculating the transition matrix, which is highly time consuming, and then obtaining its eigenvalues, also time consuming. The implicit subspace iteration method avoids calculating the transition matrix, and, instead of performing multiplications of the subspace vectors by the transition matrix, calculates their time integration along a time period. As the transition matrix is not calculated and the eigenvalue systems to solve are very small compared to the size of the transition matrix, the calculation time is reduced very significantly. If we start with any initial solution for the state over a length of time identified by the maximum time delay (τ_{max}), we can calculate the state at period T by multiplying the initial solution by the transition matrix or by performing the time integration over that period. The result of both methods must be the same, except for discrepancies given by numerical errors. Moreover, efficient time integration algorithms can result in much faster solutions, depending on the size of the state, than the time required to form the transition matrix.

The time integration along a period is a process that must be performed many times in this method. To reduce the calculation time, "step matrices" relating the state at each moment with the state one time step later are pre-calculated.

The formulation for the calculation of the step matrices in infeed grinding processes is shown next for the centerless configuration (Figure 1). In the case of cylindrical configuration, the procedure is the same just subtracting the geometrical parameters of the formulation.

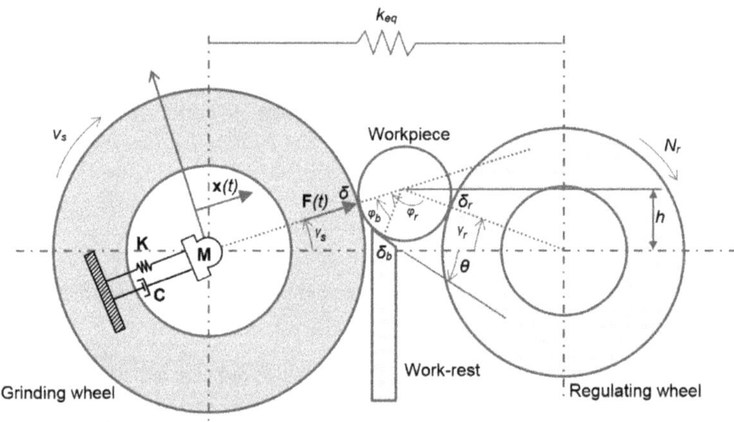

Figure 1. Centerless infeed grinding process system.

Then, the centerless infeed grinding process is defined by an Equation of motion for a multiple degree of freedom system (corresponding to considered vibration modes):

$$\mathbf{M\ddot{x}}(t) + \mathbf{C\dot{x}}(t) + \mathbf{Kx}(t) = \mathbf{F}(t) \tag{10}$$

where **M**, **C** and **K** are the mass, viscous damping and stiffness matrices of the system, and $\mathbf{F}(t)$ is the grinding force in the normal direction, which follows the next expression:

$$\mathbf{F}(t) = \phi \mathbf{P} k_{wn}[\delta r_w(t - \tau) - \delta r_w(t)] \tag{11}$$

Normalizing to unit mass, $\boldsymbol{\phi}$ is a matrix with the modal displacements of considered modes in the normal direction of the contacts between the workpiece and the grinding wheel, the regulating wheel and the support blade. k_{wn} is the cutting stiffness and $\delta r_w(t\text{-}\tau)$ and $\delta r_w(t)$ are the radius defects at the cutting point in the previous and the current workpiece revolutions, respectively, being τ the time between revolutions. $\mathbf{P}$ is a dimensionless vector relating the forces at the contact points with the normal force at the cutting point [26]. Thus, the considered direction of motion of the process system is corresponding to the normal force $\mathbf{F}(t)$, defined as $\mathbf{x}(t)$.

In centerless grinding processes, the radius defect depends on a geometric displacement of the workpiece due to roundness errors passing through the contact points with the blade and regulating wheel, as well as on the static defection of the system and the vibrations generated in the process. The radius defect at the current revolution can be expressed as:

$$\delta r_w(t) = \mathbf{C}\boldsymbol{\phi}x(t) + \frac{\mathbf{F}(t)}{k_r} + g_b\delta r_w(t - \tau_b) - g_r\delta r_w(t - \tau_r) \tag{12}$$

where k_r is the inverse of the residual flexibility. It can be calculated by subtracting the flexibility of considered vibration modes from the process equivalent stiffness k_{eq}, which is the sum of the system static flexibility and the contact flexibility between the workpiece and the grinding wheel, the regulating wheel and the support blade.

$$\frac{1}{k_r} = \frac{1}{k_{eq}} - \sum_{i=1}^{N_m} \frac{\mathbf{C}\boldsymbol{\phi}_i\boldsymbol{\phi}_i{}^t\mathbf{P}}{\omega_i^2} \tag{13}$$

where ω_i are the natural frequencies of considered modes. There are different experimental methods to obtain the equivalent stiffness [27]. $\delta r_w(t-\tau_b)$ and $\delta r_w(t-\tau_r)$ are the radius defects at the contact points with the blade and regulating wheel (represented in Figure 1 as δ_b and δ_r), and g_b and g_r are two geometrical parameters depending on the contact angles of the blade and regulating wheel with the workpiece (φ_b and φ_r), obtained from the geometric configuration of the process represented by the workpiece height h and the blade angle θ [28]. $\mathbf{C}$ is a dimensionless vector representing the displacement at the cutting point due to the displacements at the contact points [26].

Replacing Equations (11) and (12) in Equation (10) and renaming:

$$\ddot{\mathbf{x}}(t) + \mathbf{C}\dot{\mathbf{x}}(t) + [\mathbf{K} + \mathbf{V}k_t]\mathbf{x}(t) = \boldsymbol{\phi}Pk_t[\delta + g_b\delta_b - g_r\delta_r] \tag{14}$$

being $\mathbf{V} = \mathbf{C}^T\boldsymbol{\phi}\boldsymbol{\phi}^T\mathbf{P}$, $\delta_i = r_w(t - \tau)$, $\delta_b = r_w(t - \tau_b)$, $\delta_r = r_w(t - \tau_r)$ and $k_t = \dfrac{k_{wn}}{1+\left(\frac{k_{wn}}{k_r}\right)}$.

Following the same procedure of integrating and rearranging in state-space mode as [14] for cylindrical grinding process, the next expression is obtained:

$$\begin{aligned} \mathbf{p}_{i+1} = (\mathbf{A}_i - \mathbf{B}_i)\mathbf{p}_i + \mathbf{G}_{0,i}\big(\delta_{i-n} + g_b\delta_{b,i-n1} - g_r\delta_{r,i-n2}\big) \\ + \mathbf{G}_{1,i}\big(\delta_{i-n+1} + g_b\delta_{b,i-n1+1} - g_r\delta_{r,i-n2+1}\big) \end{aligned} \tag{15}$$

$$\mathbf{B}_i = \begin{bmatrix} \frac{\Delta t^2}{2} & \frac{\Delta t^3}{6} \\ \Delta t & \frac{\Delta t^2}{2} \end{bmatrix} \begin{bmatrix} \mathbf{V}k_t & 0 \\ 0 & \mathbf{V}k_t \end{bmatrix} \tag{16}$$

$$\mathbf{G}_{0,i} = \begin{bmatrix} \frac{\Delta t^2}{3} \\ \frac{\Delta t}{2} \end{bmatrix}\boldsymbol{\phi}Pk_t \,, \quad \mathbf{G}_{1,i} = \begin{bmatrix} \frac{\Delta t^2}{6} \\ \frac{\Delta t}{2} \end{bmatrix}\boldsymbol{\phi}Pk_t \tag{17}$$

Matrix $\mathbf{A}$ is constant for all the calculations. It is obtained as [21]. $\mathbf{B}_i$, $\mathbf{G}_{0,i}$ and $\mathbf{G}_{1,i}$ vary and they are calculated for each discretized segment. This set of matrices represents the step matrices of the process.

Once all the step matrices are calculated, the procedure to obtain the state after one period begins with the definition of the initial state of the process by including the initial discretized radius defect and the initial dynamic modal solution, presented in the following mixed system state:

$$\mathbf{q}_n = \left\{ \begin{matrix} \delta_1 & \delta_2 & \cdots & \delta_{n-2} & \delta_{n-1} & x_n & \dot{x}_n \end{matrix} \right\} \tag{18}$$

For the first segment:

$$\mathbf{p}_1 = \mathbf{A}\mathbf{p}_n \tag{19}$$

$$\delta_1 = \mathbf{T}\mathbf{p}_1 + W\delta_{1-n} + G_b\delta_{1-n1} - G_r\delta_{1-n2} \tag{20}$$

$$\mathbf{T} = \frac{\mathbf{C}^T \boldsymbol{\phi}}{\left(1 + \frac{k_{wn}}{k_r}\right)}, \quad W = \frac{k_t}{k_r} \tag{21}$$

Reaching the last segment:

$$\delta_{i+1} = \mathbf{T}\mathbf{p}_{i+1} + W\delta_{i+1-n} + G_b\delta_{i+1-n1} - G_r\delta_{i+1-n2} \tag{22}$$

$$\begin{aligned} \mathbf{p}_{i+1} = (\mathbf{A}_i - \mathbf{B}_i)\mathbf{p}_i + \mathbf{G}_{0,i}\left(\delta_{i-n} + g_b\delta_{b,i-n1} - g_r\delta_{r,i-n2}\right) \\ + \mathbf{G}_{1,i}\left(\delta_{i-n+1} + g_b\delta_{b,i-n1+1} - g_r\delta_{r,i-n2+1}\right) \end{aligned} \tag{23}$$

The calculation of the product of the complete state vector using the step matrices, as proposed here, leads to a lower requirement of mathematical operations compared to the product by the transition matrix. Besides, the calculation of the transition matrix is the operation which requires the highest computational effort compared to the calculation of the step matrices when obtaining the stability maps for grinding processes.

Another important issue is the calculation time dependence on the size of the state, represented by the discretization segments. The requirement is that the integration time, equal to the length of the segment divided by the workpiece rotational speed, must be below 0.1 times the corresponding period related to the maximum natural frequency considered.

The criterion of stability is the same as the one used in the semi-discretization method. Then, if any eigenvalue of the relation between the initial and final states has a magnitude larger than 1, then the system is unstable, whereas when all magnitudes are lower than 1, the system is stable.

Besides dynamic instabilities, there are also geometric instabilities in centerless grinding processes due to their special geometric configuration [26], named geometric lobing. They are included in the model by the geometric parameters corresponding to the contact points between the workpiece and the regulating wheel and the blade support. For that reason, it is necessary to include different lobe numbers in the initial radius defect of the process. Then, the evolution of this lobe for each geometric configuration can be analyzed and included in the stability diagrams as a result of the subspace iteration method. Cylindrical infeed grinding processes do not give rise to that phenomenon.

As in the semi-discretization method in [14], with the developed method to analyze the stability in grinding processes there is also the possibility to include the continuous workpiece speed variation in the calculation.

The condition is that the speed variation period is an integer multiple of the workpiece speed period. Then, the calculation of the step matrices must be done over the speed variation period and the eigenvectors analysis is carried out between the first and the last system states.

Therefore, the same stability maps presented in [14,19] can be obtained, whose axes are the amplitude and frequency of the workpiece speed variation.

3. Results

3.1. Application of the Method to Cylindrical Grinding Processes

First application of the method is carried out for cylindrical infeed grinding process by comparing the results with the data proposed in Alvarez et al. [14]. In this analysis, a validation of the accuracy of the new method was done and a comparison of the computational effort of both semidiscretization and subspace iteration methods is presented.

The process conditions for stability analysis in [14] are shown in Table 1.

Table 1. Process conditions for stability analysis in cylindrical infeed grinding [14].

Modal Analysis			
Mode	f_n **(Hz)**	ξ **(%)**	Φx
1	200	5	0.1
2	300	5	0.09
3	400	4	0.1
Grinding wheel diameter (mm)		600	
Grinding wheel speed (m/s)		50	
Workpiece diameter (mm)		25	
Equivalent stiffness (N/μm)		50	
Specific energy (J/mm^3)		40	

Figure 2a shows the stability map obtained with the subspace iteration method for the process conditions of Table 1, which represents the stability degree of each combination of workpiece speed and ground length. Values below the stability limit of 1 are stable combinations and values over the stability limit are unstable combinations. Almost the same stability map as the one obtained with the semi-discretization method is achieved. Figure 2b shows isolines with different degrees of stability. Dash lines correspond to subspace iteration calculations, while solid lines correspond to semidiscretization calculations. Good correlation between isoline values can be noticed.

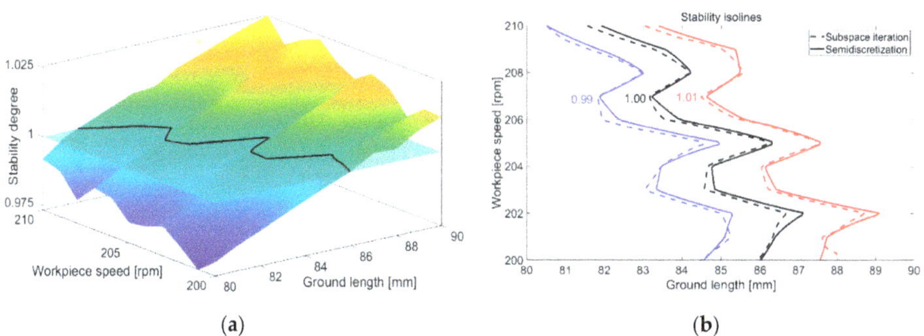

Figure 2. (a) Subspace iteration stability map; (b) comparison of stability isolines between semi-discretization and subspace iteration methods.

Necessary computational effort for achieving the stability map has been reduced significantly. The time required for obtaining the map with the semi-discretization method is 11,441 s, while the time with the subspace iteration method is 256 s (45 times faster), considering that the map corresponds to a 11×11 matrix of stability values. For this comparison, an Intel $^{®}$ Core ^{TM} i7-4610M CPU 3.00 GHz with RAM 16 GB has been used.

One important aspect is the dependency of the number of discretized points on the calculation time. As mentioned above, the integration time is calculated based on the considered maximum natural frequency. Figure 3 shows the comparison of the calculation time of both semi-discretization and subspace iteration methods related to that frequency. As it can be noticed, the higher the frequency the higher the difference between both calculations. For a natural frequency of 100 Hz, the subspace iteration method is six times faster, being 138 times faster for a frequency of 700 Hz. Therefore, the subspace iteration method is even more suitable for systems with higher natural frequencies.

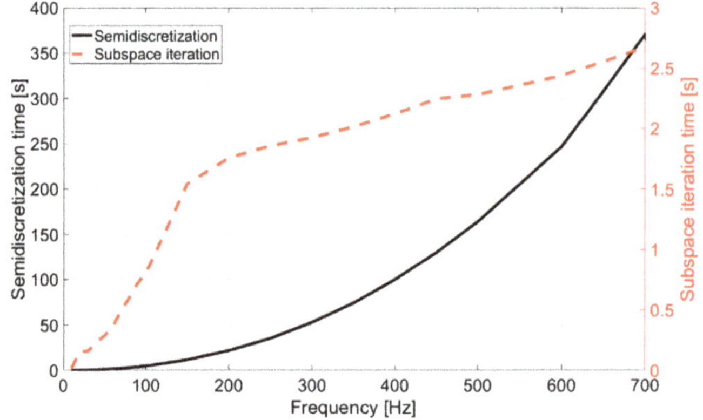

Figure 3. Comparison of time for calculation of stability maps between semi-discretization and subspace iteration methods for cylindrical grinding.

The subspace iteration method is also valid to calculate the stability maps with continuous workpiece speed variations, whose axes are the frequency and amplitude of the sinusoidal variation. Figure 4 shows a comparison of the method with that of the semi-discretization method following the same procedure. Stability maps are calculated with process conditions of [14].

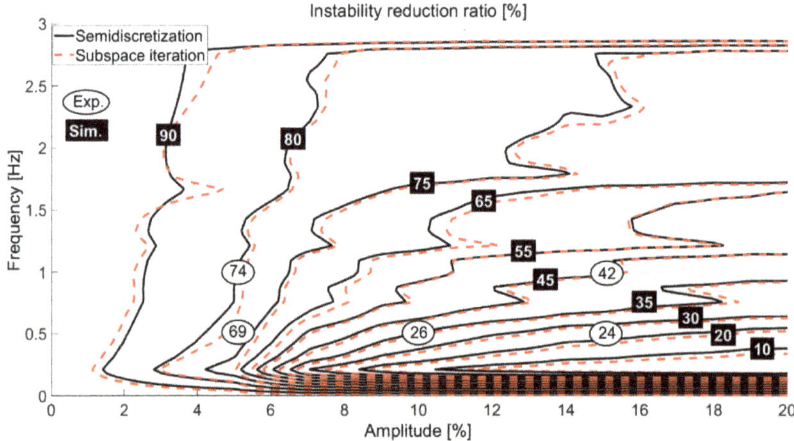

Figure 4. Comparison of instability reduction ratio maps with the application of continuous workpiece speed variation, obtained with both methods.

Isolines of instability reduction ratio are presented for both methods and good correlation can be noticed. Black boxes show the theoretical values of isolines while white ellipses show the experimental values.

The time required for obtaining the map with the semi-discretization method is 46,314 s, while the time with the subspace iteration method is 2773 s (16 times faster), considering that the map corresponds to a 16×26 matrix of instability reduction ratio values.

Calculation time for achieving this type of map is increased due to the necessity of calculating the step matrices over the speed variation period. Therefore, the higher the speed variation frequency, the lower the calculation time. Figure 5 shows the time required for frequencies of 0.5 and 3 Hz, where the matrices must be recalculated six and two times, respectively. In this case, for a natural frequency of 100 Hz, the subspace iteration method is 2.5 times faster, being 47 times faster for a frequency of 700 Hz.

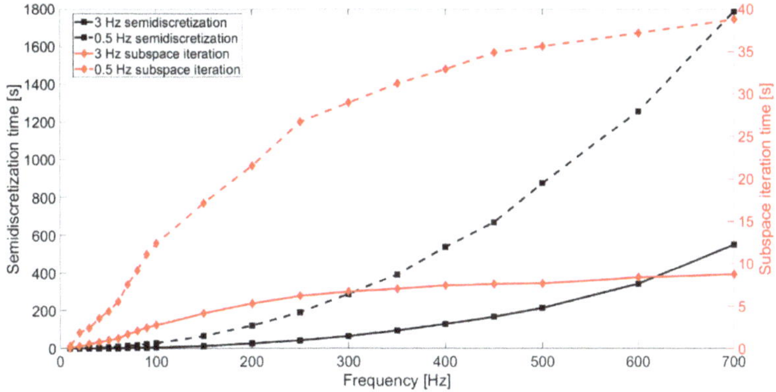

Figure 5. Comparison of time for calculation of stability maps between semi-discretization and subspace iteration methods for cylindrical grinding with the application of continuous workpiece speed variation.

3.2. Application of the Method to Centerless Grinding Processes

Firstly, the application of the subspace iteration method to centerless grinding processes is validated by comparing it to the work by Barrenetxea et al. [23], where stability maps are calculated and validated in the frequency domain. Figure 6a shows the map obtained in the frequency domain and Figure 6b shows the one by the subspace iteration method. Good correlation between both can be noticed. The main difference is that the stability limit for frequency domain is equal to zero with negative unstable conditions, while subspace iteration is equal to 1 with unstable conditions over that value.

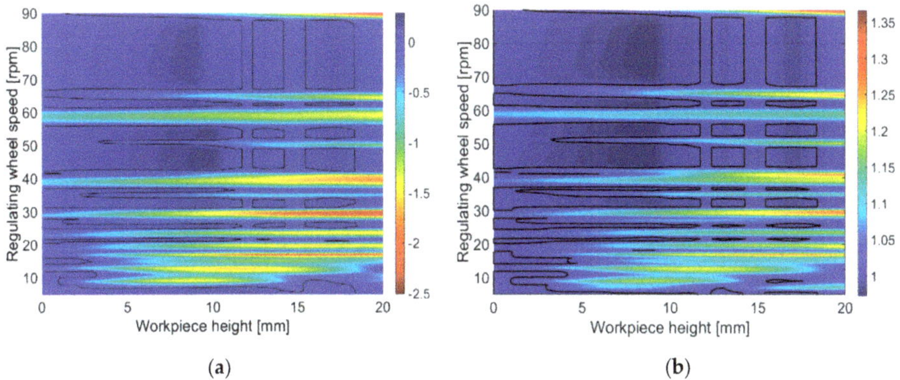

Figure 6. Dynamic stability map for centerless grinding by (**a**) frequency domain method [21]; (**b**) subspace iteration method.

In the case of centerless grinding processes, aside from the work done by Barrenetxea et al. [19], there is no bibliography regarding the obtainment of stability maps with continuous workpiece speed variation. Then, the subspace iteration method is compared to the time domain simulation procedure by Barrenetxea et al. [19] in terms of calculation accuracy and computational time consumption. Figure 7 shows the map obtained in [19] by the time simulation approach while Figure 8 shows the one obtained by the subspace iteration method. Good correlation can be noticed again, validating the accuracy of the proposed method.

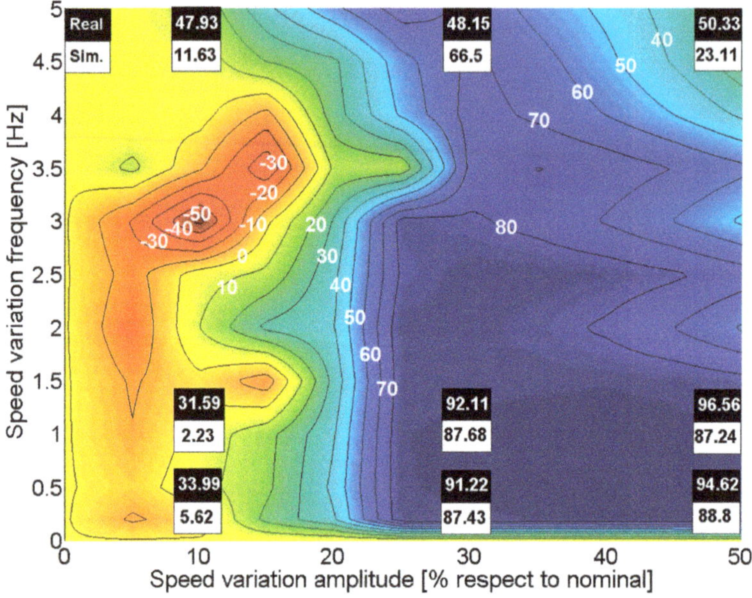

Figure 7. Instability reduction ratio for centerless grinding by time simulation method [19].

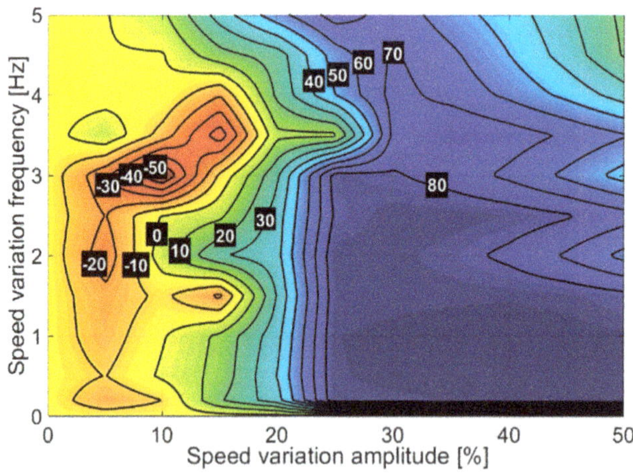

Figure 8. Instability reduction ratio for centerless grinding by subspace iteration method.

Calculation time for achieving this type of instability reduction ratio in centerless grinding with the application of continuous workpiece speed variation is represented in the Figure 9. In this case, the time-domain simulations are independent of the variation frequency, since the simulation is done during all the workpiece revolutions of the whole grinding cycle. The time required for achieving the time-domain 11 × 11 map of Figure 7 is 63,525 s, while the time for achieving the subspace iteration map of Figure 8 is 2541 s, being 25 times faster.

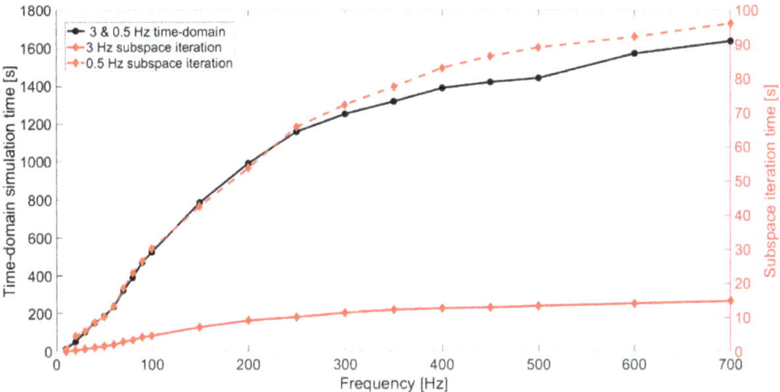

Figure 9. Comparison of time for calculation of stability maps between time-domain and subspace iteration methods for centerless grinding with the application of continuous workpiece speed variation.

3.3. Conclusions

The results obtained from the comparison between subspace iteration and semi-discretization methods in cylindrical grinding process show that:

- When obtaining stability maps, subspace iteration is between 6 and 138 times faster than semi-discretization in the analized cases, depending on the natural frequencies of the system.
- Both methods are dependent on the natural frequencies of the system, but subspace iteration is better when dealing with higher frequencies.
- When obtaining instability reduction ratio maps with continuous workpiece speed variation, subspace iteration is between 2.5 and 47 times faster than semi-discretization in the analyzed cases, depending on the natural frequencies of the system.
- The time necessary for the calculation of instability reduction ratio maps is higher than that required for the calculation of stability maps due to the necessity of calculating the step matrices over the speed variation period.

Regarding the results obtained comparing subspace iteration and time simulation methods in centerless grinding process:

- When obtaining instability reduction ratio maps with continuous workpiece speed variation, subspace iteration is about 27 times faster than time simulation in the analized case.
- Since time domain simulation does not depend on the workpiece speed variation frequency, the improvement of the subspace iteration calculation time is higher when analyzing higher variation frequencies.

4. Discussion

The subspace iteration method has been applied successfully to the generation of stability maps for cylindrical and centerless infeed grinding processes. The method has also been validated to calculate the instability reduction ratio maps with the application of continuous workpiece speed variation in cylindrical and centerless infeed grinding processes.

Accuracy of this method has been validated by comparing it with previously published methods such as semi-discretization, frequency and time-domain simulations, obtaining almost the same results and validating the subspace iteration method in grinding processes.

Reduction of calculation time related to the semi-discretization method is achieved in analyzed cases because the transition matrix is not explicitly calculated and, instead of calculating all the eigenvectors of the transition matrix, an iterative process calculating only the dominant eigenvalues is used. In the comparison with the semi-discretization method for cylindrical infeed grinding, subspace iteration is up to 138 times faster for stability maps and 47 times faster for instability reduction ratio maps. In the comparison with time-domain for centerless infeed grinding, subspace iteration is 27 times faster for instability reduction ratio maps.

The influence of the number of segments related to the maximum natural frequency of the grinding system has been analyzed, concluding that subspace iteration has a better performance for a higher number of segments than semi-discretization.

Author Contributions: Conceptualization, J.A. and M.Z.; Data curation, J.A. and B.I.; Formal analysis, M.Z. and J.I.M.; Investigation, J.A. and M.Z.; Methodology, D.B. and J.I.M.; Project administration, D.B.; Resources, J.A., D.B. and J.I.M.; Software, J.A.; Supervision, M.Z.; Validation, J.A., D.B. and B.I.; Visualization, J.A. and B.I.; Writing—original draft, J.A. and D.B.; Writing—review & editing, J.A. and M.Z. All authors have read and agreed to the published version of the manuscript.

Funding: This research received no external funding.

Conflicts of Interest: The authors declare no conflict of interest.

References

1. Marinescu, I.D.; Hitchiner, M.; Uhlmann, E.; Rowe, W.B.; Inasaki, I. *Handbook of Machining with Grinding Wheels*; CRC Press: Boca Raton, FL, USA, 2006.
2. Inasaki, I.; Karpuschewski, B.; Lee, H. Grinding chatter—Origin and suppression. *CIRP Ann. Manuf. Tech.* **2001**, *50*, 515–534. [CrossRef]
3. Hahn, R.S. On the Theory of Regenerative Chatter in Precision Grinding Operations. *Trans. ASME* **1954**, *76*, 593–597.
4. Snoeys, R.; Brown, D. Dominating Parameters in Grinding Wheel and Workpiece Regenerative Chatter. In Proceedings of the 10th International Machine Tool Design and Research Conference, Manchester, UK, 17 September 1969; pp. 325–348.
5. Thompson, R.A. On the Doubly Regenerative Stability of a Grinder: The Combined Effect of Wheel and Workpiece Speed. *ASME J. Eng. Ind.* **1977**, *99*, 237–241. [CrossRef]
6. Thompson, R.A. On the Doubly Regenerative Stability of a Grinder: The Theory of Chatter Growth. *ASME J. Eng. Ind.* **1986**, *108*, 75–82. [CrossRef]
7. Liao, Y.S.; Shiang, L.C. Computer Simulation of Self-Excited and Forced Vibrations in the External Cylindrical Plunge Grinding Process. *ASME J. Eng. Ind.* **1991**, *113*, 297–304. [CrossRef]
8. Li, H.; Shin, Y.C. A Time-Domain Dynamic Model for Chatter Prediction of Cylindrical Plunge Grinding Processes. *J. Manuf. Sci. Eng.* **2006**, *128*, 404–415. [CrossRef]
9. Biera, J.; Vinolas, J.; Nieto, F.J. Time Domain Dynamic Modeling of the External Plunge Grinding Process. *Int. J. Mach. Tools Manuf.* **1997**, *37*, 1555–1572. [CrossRef]
10. Chung, K.W.; Liu, Z. Nonlinear analysis of chatter vibration in a cylindrical transverse grinding process with two time delays using a nonlinear time transformation method. *Nonlinear Dyn.* **2011**, *66*, 441–456. [CrossRef]
11. Insperger, T.; Mann, B.; Stépán, G.; Bayly, P. Stability of up-milling and down-milling, part 1: Alternative analytical methods. *Int. J. Mac.h Tools Manuf.* **2003**, *43*, 25–34. [CrossRef]
12. Insperger, T.; Stépán, G. Stability analysis of turning with periodic spindle speed modulation via semidiscretization. *J. Vib. Control* **2004**, *10*, 1835–1855. [CrossRef]
13. Iglesias, A.; Dombovari, Z.; Gonzalez, G.; Munoa, J.; Stepan, G. Optimum Selection of Variable Pitch for Chatter Suppression in Face Milling Operations. *Materials* **2019**, *12*, 112. [CrossRef] [PubMed]

14. Alvarez, J.; Zatarain, M.; Barrenetxea, D.; Ortega, N.; Gallego, I. Semi-discretization for Stability Analysis of In-feed Cylindrical Grinding with Continuous Workpiece Speed Variation. *Int. J. Adv. Manuf. Technol.* **2013**, *69*, 113–120. [CrossRef]
15. Alvarez, J.; Zatarain, M.; Marquinez, J.I.; Ortega, N.; Gallego, I. Avoiding Chatter in Traverse Cylindrical Grinding by Continuous Workpiece Speed Variation. *J. Manuf. Sci. Eng.* **2013**, *135*, 051011. [CrossRef]
16. Zatarain, M.; Bediaga, I.; Muñoa, J.; Lizarralde, R. Stability of Milling Processes with Continuous Spindle Speed Variation: Analysis in the Frequency and Time Domains, and Experimental Correlation. *CIRP Ann. Manuf. Tech.* **2008**, *57*, 379–384. [CrossRef]
17. Merino, R.; Bediaga, I.; Iglesias, A.; Munoa, J. Hybrid Edge—Cloud-Based Smart System for Chatter Suppression in Train Wheel Repair. *Appl. Sci.* **2019**, *9*, 4283. [CrossRef]
18. Inasaki, I.; Cheng, C.; Yonetsu, S. Suppression of chatter in grinding. *Bull. Jpn. Soc. Precis. Eng.* **1976**, *9*, 133–138.
19. Barrenetxea, D.; Marquinez, J.I.; Bediaga, I.; Uriarte, L. Continuous Workpiece Speed Variation (CWSV): Model based practical application to avoid chatter in grinding. *CIRP Ann. Manuf. Tech.* **2009**, *58*, 319–322. [CrossRef]
20. Alvarez, J.; Barrenetxea, D.; Marquinez, J.I.; Bediaga, I.; Gallego, I. Effectiveness of continuous workpiece speed variation (CWSV) for chatter avoidance in throughfeed centerless grinding. *Int. J. Mach. Tools Manuf.* **2011**, *51*, 911–917. [CrossRef]
21. Zatarain, M.; Alvarez, J.; Bediaga, I.; Munoa, J.; Dombovari, Z. Implicit Subspace Iteration as an Efficient Method to Compute Milling Stability Lobe Diagrams. *Int. J. Adv. Manuf. Technol.* **2015**, *77*, 597–607. [CrossRef]
22. Zatarain, M.; Dombovari, Z. Stability analysis of milling with irregular pitch tools by the implicit subspace iteration method. *Int. J. Dyn. Control* **2014**, *2*, 26–34. [CrossRef]
23. Barrenetxea, D.; Alvarez, J.; Marquinez, J.I.; Gallego, I.; Muguerza, I.; Krajnik, P. Stability analysis and optimization algorithms for the set-up of infeed centerless grinding. *Int. J. Mach. Tools Manuf.* **2014**, *84*, 17–32. [CrossRef]
24. Floquet, M.G. Équations différentielles linéaires à coefficients périodiques. *Ann. Sci. École Norm. Supérieure* **1883**, *8*, 1–36.
25. Dombovari, Z.; Iglesias, A.; Zatarain, M.; Insperger, T. Prediction of multiple dominant chatter frequencies in milling processes. *Int. J. Mach. Tools Manuf.* **2011**, *51*, 457–464. [CrossRef]
26. Gallego, I. Intelligent Centerless Grinding: Global Solution for Process Instabilities and Optimal Cycle Design. *CIRP Ann.* **2007**, *56*, 347–352. [CrossRef]
27. Ramos, J. Caracterización del Comportamiento Dinámico de Máquinas-Herramienta. Aplicación al Rectificado Cilíndrico de Exteriores en Penetración y al Fresado Frontal Vertical. Ph.D. Thesis, Universidad de Navarra, Pamplona, Spain, 1998.
28. Lizarralde, R.; Barrenetxea, D.; Gallego, I.; Marquinez, J.I.; Bueno, R. Practical Application of New Simulation Methods for the Elimination of Geometric Instabilities in Centerless Grinding. *CIRP Ann.* **2005**, *54*, 273–276. [CrossRef]

Publisher's Note: MDPI stays neutral with regard to jurisdictional claims in published maps and institutional affiliations.

Article

Uncharted Stable Peninsula for Multivariable Milling Tools by High-Order Homotopy Perturbation Method

Jose de la Luz Sosa [1], Daniel Olvera-Trejo [1,*], Gorka Urbikain [2,*], Oscar Martinez-Romero [1], Alex Elías-Zúñiga [1] and Luis Norberto López de Lacalle [2]

[1] Tecnológico de Monterrey, Escuela de Ingeniería y Ciencias, Av. Eugenio Garza Sada 2501, Monterrey, Nuevo León 64849, Mexico; A00828123@itesm.mx (J.d.l.L.S.); oscar.martinez@tec.mx (O.M.-R.); aelias@tec.mx (A.E.-Z.)

[2] Department of Mechanical Engineering, University of the Basque Country, Alameda de Urquijo s/n, 48013 Bilbao, Spain; norberto.lzlacalle@ehu.eus

* Correspondence: daniel.olvera.trejo@tec.mx (D.O.-T.); gorka.urbikain@ehu.eus (G.U.)

Received: 9 October 2020; Accepted: 3 November 2020; Published: 6 November 2020

Abstract: In this work, a new method for solving a delay differential equation (DDE) with multiple delays is presented by using second- and third-order polynomials to approximate the delayed terms using the enhanced homotopy perturbation method (EMHPM). To study the proposed method performance in terms of convergency and computational cost in comparison with the first-order EMHPM, semi-discretization and full-discretization methods, a delay differential equation that model the cutting milling operation process was used. To further assess the accuracy of the proposed method, a milling process with a multivariable cutter is examined in order to find the stability boundaries. Then, theoretical predictions are computed from the corresponding DDE finding uncharted stable zones at high axial depths of cut. Time-domain simulations based on continuous wavelet transform (CWT) scalograms, power spectral density (PSD) charts and Poincaré maps (PM) were employed to validate the stability lobes found by using the third-order EMHPM for the multivariable tool.

Keywords: chatter; multivariable tool; stable peninsula; homotopy perturbation method

---

1. Introduction

There are many phenomena in different fields of science and engineering where the physical response of a variable involves not only the value at time t but also the effects that occur in an earlier state $t - \tau$. Thus, delay systems appear in many engineering problems, such as in the shimmy effect (wheel vibration) [1], vehicle traffic models [2], feedback stabilization problems [3] and in the regenerative vibration of machine-tools better known as chatter [4]. In cases where the net force depends on the current values and some past values (history) such as position and speed, the system dynamic behavior can be modeled using a differential delay equation (DDE).

It is well-known that during a milling process, unstable vibrations also known as self-excited vibration or chatter may occur. Chatter reduces the machining efficiency due to low material removal rate by reducing the workload and affects surface quality, shortens tool life and accelerates tool wear. Researchers are studying several ways to overcome this limitation. Kuljanic et al. [5] studied the incorporation of a chatter detection system based on multiple sensors to milling operations for industrial conditions, Zhuo et al. [6] used a method based on fractal dimension for the flank milling of a thin-walled blade, which can reflect the chatter severity level through the morphological change in signal. Paul and Morales [7], to mitigate chatter, presented an active controller based on the technique of discrete time sliding mode control (DSMC) blended with the type-2 fuzzy logic system. Moreover, Peng et al. [8] presented a method based on a dynamic cutting force simulation model and a machine learning approach based on statistical learning theory to predict and avoid the cutting chatter.

In addition, to control and suppress chatter vibrations, the use of piezoelectric actuators embedded in the tool holder [9], electromagnetic actuators integrated into the spindle system [10] and tunable clamping table [11] has been analyzed. In the milling process, the use of variable pitch cutters has demonstrated to improve productivity [12]. Different from the uniform pitch cutter, when a variable pitch cutter is used the dynamics model of cutting vibration changes from DDEs with a single delay to DDEs with multiple delays [13]. A common technique offline to predict unstable vibrations is the so-called stability lobes of the DDE based on Floquet theory [14], in which a curve describes the limit of stable vibration under feasible range values of cutting parameters.

The stability analysis of the milling process with multiple delays has been studied through different methods. Among all these methods, those with variable pitch tools play a critically important role [15]. Slavicek [16] was the one who first demonstrated the effectiveness of variable pitch cutters in suppressing vibrations in the milling process, he assumed a rectilinear tool motion for cutting teeth, and applied the theory of orthogonal stability to the irregular pitch of the tooth, by assuming an alternating step variation then, he obtained an expression of the stability limit as a function of the step angle variation. Budak [17,18] proposed an analytical method for nonconstant pitch milling cutters from a design point of view, showing for some applications how this variable effect helps to reduce self-excited vibrations, so he found that chatter stability can be improved significantly even at slow cutting speeds by properly designing the pitch angles. Altintas et al. [19] used the frequency domain method to analyze the milling stability of the variable pitch cutter and introduced a method to select the optimal pitch angles. Olgac and Sipahi proposed a mathematical approach, the cluster treatment of characteristic roots (CTCR), which optimizes the design of variable pitch cutters [20]. Jin et al. [21] presented an improved semi-discretization algorithm to predict the stability lobes for variable pitch cutters, which were verified and compared with previous works such as the Altintas analytical method (zero-order method) [19]. Comak and Budak [22] showed the optimal design of a tool for milling operations with variable geometry to widen the stability zones using the semi-discretization method, validating it experimentally. They also used a design methodology to determine the optimal pitch angle geometry for a given cutting condition, allowing increased stability.

Zatarain et al. [23] extended the multifrequency solution proposed by Budak and Altintas [24] to include the helix effect, they pointed out that the variation of the helix angle plays an important role in stability graphs due to repetitive vibrations driven by impact (flip), they found that the flip lobes became closed curves that are separated by horizontal lines where the depth of cut is equal to a multiple of the helix pitch. A similar phenomenon was confirmed using the semi-discretization method (SDM) in [25], meanwhile, B.R. Patel et al. [26] considered the influence of the helix angle of the tool to obtain an analytical force model, they found that isolated islands of instability can occur in the milling processes, which are induced by the helix angle of the tool and lead to separate regions of period-doubling and quasi-period behavior. Sims et al. [27] by using an adapted and time-averaged version of the SDM analyzed both the influence on the variation of the helix angle and the pitch angle of the tool to improve the prediction of vibrations and estimate predictions of surface errors. They used the semi-discretization method, the time-averaged semi-discretization method and the temporal finite element method to predict vibration stability for variable helix and variable pitch milling tools. Turner et al. [28] modeled and compared stability for variable pitch and helix angle cutters, demonstrating that variable helix angle tools can have higher stability and productivity.

Yusoff and Sims in [29] combined SDM with differential evolution to optimize variable helix milling tools to minimize vibration, their analysis predicted total vibration mitigation using the optimized variable helix milling tool at low radial immersion. Furthermore, Dombovari and Stepan [30] introduced a general mechanical model based on SDM to predict the linear stability of specialty cutters with optional continuous variation of the helix angle. Using an extended second-order SDM, Zhan et al. [15] predicted the stability lobe diagrams for tools with variable pitch angles. Meanwhile, Huang et al. [31] conducted a stability analysis for milling operations with variable pitch mills at variable speed, while Cai et al. [32] proposed an integrated process machine model based on the computer graphics method to simulate the milling process of a variable pitch cutter.

On the other hand, Olvera and Elías-Zuñiga in [33] led to the development of the enhanced multistage homotopy perturbation method (EMHPM) to solve differential delay equations (DDEs) with constant and variable coefficients and then this EMHPM was applied to predict the stability of a multivariate milling tool in which they consider the helix angle and the pitch angle variation of the cutting edges [34]. Based on the Laplace formulation, Sims [35] studied the stability of milling operations with a variable helix angle. Using the multi-frequency solution, Otto et al. [36] derived a dynamic process model where the non-linear shear force and the runout effect are included for milling with non-uniform pitch and variable helix tools. Niu et al. [37] found that runout can significantly increase the stability limits regardless of spindle speed ranges, while Olvera et al. [38] in a study for a thin-walled workpiece demonstrated that by considering the effects of the runout, the helix angle and characterization dependent on the cutting speed, more precise stability boundaries are achieved.

To demonstrate that one of the effective ways to suppress vibration in milling operations is to use tools with variable pitch and helix angle, Wang et al. [12] proposed an improved semi-discretization method based on Floquet's theory. Since the delay between each cutting edge varies along with the axial depth of the tool in milling, they discretized the cutting tool in some axial layers to simplify the calculation. Iglesias et al. [39] presented a method to find the optimal angles between the inserts, and the stability diagrams were obtained through the iterative brute force (BF) method, which consists of an iterative maximization of stability through the semi-discretization method. They conclude that, if an optimal selection of the angle between the inserts is possible then, the material removal rate can be improved up to three times. Gou et al. [40] proposed an effective optimization method for the variable helical cutter introducing an index called "suppression factor" to measure stability quantitatively.

Therefore, in the present work, the EMHPM developed in [33] and extended for analysis of multivariable tools in [34], is now expanded to solve the dynamics of the machining process in milling in which the approximation to the delay is performed with polynomials of degree two and three. In order to study the proposed method performance in terms of convergence and computational cost, a multivariable milling tool with a variable pitch cutter and helix angle is used to determine milling process in stability domains.

This paper is summarized as follows. Section 2 focuses on the development of second- and third-order EMHPM for stability analysis of DDE. Section 3 studies the application of the second- and third-order EMHPM on the milling equation to demonstrate its improvement in the convergence rate. Section 4 is focused on the use of the third-order EMHPM to compute the stability analysis in milling for multivariable tools, and theoretical predictions with time-domain simulations are performed. Finally, some conclusions are drawn.

2. Enhanced Multistage Homotopy Perturbation Method

2.1. Second-Order EMHPM

Olvera et al. enhanced in [33] the multistage homotopy perturbation method (MHPM) proposed by Hashim [41]. The EMHPM considers the general case in which the nonlinear equation contains terms of the independent variable. This method is also useful to solve an n-dimensional DDE in the state-space form

$$\dot{\mathbf{x}}(t) = \mathbf{A}(t)\mathbf{x} + \mathbf{B}(t)\mathbf{x}(t - \tau) \tag{1}$$

where $\mathbf{A}(t + \tau) = \mathbf{A}(t)$, $\mathbf{B}(t + \tau) = \mathbf{B}(t)$, $\mathbf{x}(t)$, is the state vector, and τ is the time delay. Equation (1) can be written equivalently as:

$$\dot{\mathbf{x}}_i(T) - \mathbf{A}_t\mathbf{x}_i(T) \approx \mathbf{B}_t\mathbf{x}_i^\tau(T) \tag{2}$$

where $\mathbf{x}_i(T)$ indicates the m-order solution for Equation (1) that satisfies the initial conditions $\mathbf{x}_i(0) = \mathbf{x}_{i-1}$, $\mathbf{A}_t$ and $\mathbf{B}_t$ are the periodic matrix whose values vary with time t. In [42], Puma et al. applied the first-order EMHPM to estimate the delayed term $x_i^{\tau}(T)$ in Equation (2), in which the period $[t_0 - \tau, t_0]$ was discretized in N equally spaced discrete state values, and the function that describes the delayed term $x_i^{\tau}(T)$ in the delayed interval $[t_{i-N}, t_{i-N+1}]$ was approximated as a first-order polynomial representation. Defining $x_i \equiv x_i(T)$ to simplify the notation, Equation (2) can be written as

$$\dot{x}_i(T) = \mathbf{A}_t x_i(T) + \mathbf{B}_t\left(x_{i-N} + \frac{N-1}{\tau}(x_{i-N+1} - x_{i-N})T\right) \tag{3}$$

Figure 1a shows the representation of the approximation of the delayed term with the first-order polynomial. In the second-order EMHPM, to approximate the function that describes the delayed term $x_i^{\tau}(T)$ in Equation (2), the Lagrange equation is used, making use of the discrete values x_{i-N}, x_{i-N+1}, x_{i-N+2} as follows:

$$f_n(x) = \sum_{i=0}^{n} Li(x)f(x_i), \; L_i(x) = \prod_{i=0, i \neq k}^{n} \frac{x - x_i}{x_k - x_i} \tag{4}$$

to achieve a second-degree polynomial approximation, we have from Equation (4) that

$$P_2(x) = \frac{(x - \Delta t)(x - 2\Delta t)}{(0 - \Delta t)(0 - 2\Delta t)}f(x_{i-N}) + \frac{(x - 0)(x - 2\Delta t)}{(\Delta t - 0)(\Delta t - 2\Delta t)}f(x_{i-N+1}) + \frac{(x - 0)(x - \Delta t)}{(2\Delta t - 0)(2\Delta t - \Delta t)}f(x_{i-N+2}) \tag{5}$$

(a)   (b)

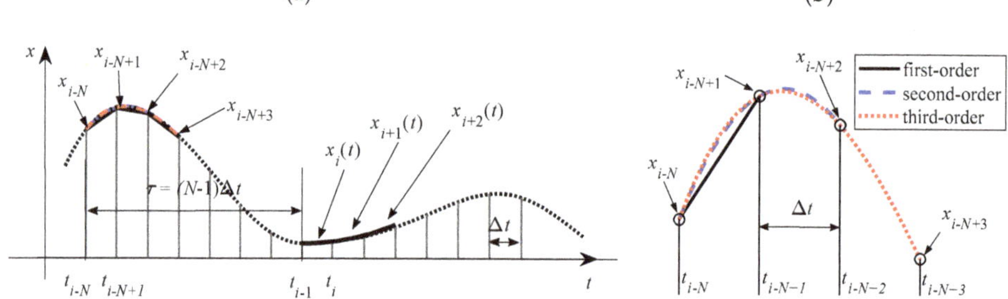

Figure 1. (**a**) Scheme for the approximation of the delayed term by a first-order (solid black line), second-order (dashed blue line) and third-order (dotted red line) polynomial: (**b**) zoom in the time interval $[t_{i-N+1}, t_{i-N+3}]$.

Substituting $x = T$ and $\Delta t = (N-1)/\tau$, we obtain the function that describes the delayed interval as:

$$x_{i-N+1}(T) \approx x_{i-N} + \left(\frac{N-1}{\tau}\right)T\left(-\frac{3}{2}x_{i-N} + 2x_{i-N+1} - \frac{1}{2}x_{i-N+2}\right) + \left(\frac{N-1}{\tau}\right)^2\frac{T^2}{2}(x_{i-N} - 2x_{i-N+1} + x_{i-N+2}) \tag{6}$$

When the delay is approximated by a second-degree polynomial it is called second-order EMHPM and should not be confused with the order of solution m and which is determined by the last deformation taking into account the approximated solution. Notice that a polynomial of the second-degree requires three points. Likewise four points in the case of a third-degree polynomial as shown in Figure 1.

The procedure to calculate the second-order EMHPM solution is based on the EMHPM procedure described in [33]. The solution for second-order EMHPM is recursively expressed of $X_{ik}(T)$ as

$$X_{ik} = X_{ik}^a + X_{ik}^b + X_{ik}^c, k = 1, 2, 3 \ldots. \tag{7}$$

where

$$X_{i0}^a = x_{i-1}, \; X_{i0}^b = X_{i0}^c = 0 \tag{8}$$

and

$$\mathbf{X}_{ik}^{a} = \tfrac{T}{k}\left(\mathbf{A}_t \mathbf{X}_{i(k-1)}^{a} + g(k)\mathbf{B}_t \mathbf{x}_{i-N}\right)$$

$$\mathbf{X}_{ik}^{b} = \tfrac{T}{k+1}\left(\mathbf{A}_t \mathbf{X}_{i(k-1)}^{b} + g(k)\left(\tfrac{N-1}{\tau}\right)\mathbf{B}_t T\left(-\tfrac{3}{2}\mathbf{x}_{i-N} + 2\mathbf{x}_{i-N+1} - \tfrac{1}{2}\mathbf{x}_{i-N+2}\right)\right)$$

$$\mathbf{X}_{ik}^{c} = \tfrac{T}{k+2}\left(\mathbf{A}_t \mathbf{X}_{i(k-1)}^{c} + g(k)\left(\tfrac{N-1}{\tau}\right)^{2}\mathbf{B}_t \tfrac{T^2}{2}\left(\mathbf{x}_{i-N} - 2\mathbf{x}_{i-N+1} + \mathbf{x}_{i-N+2}\right)\right) \tag{9}$$

So, the solution of Equation (1) is obtained by adding each of the approximations $\mathbf{X}_{ik}$ of Equation (7).

$$\mathbf{x}_i(T) \approx \sum_{k=0}^{m} \mathbf{X}_{ik}(T) \tag{10}$$

2.2. Third-Order EMHPM Solution

For the polynomial representation of the third-degree, the function that describes the delayed term $\mathbf{x}_i^{\tau}(T)$ is approximated by a polynomial of order three, then Equation (4) of the Lagrange interpolator is used accordingly. In this case, it is necessary to employ the $\mathbf{x}_{i-N}$, $\mathbf{x}_{i-N+1}$, $\mathbf{x}_{i-N+2}$, $\mathbf{x}_{i-N+3}$ discrete values. Following the same procedure described in Section 2.1, the function that describes the delayed interval is given as:

$$\mathbf{x}_i^{\tau}(T) = \; \mathbf{x}_{i-N+1}(T) \approx \mathbf{x}_{i-N} + \left(\tfrac{N-1}{\tau}\right)T\left(-\tfrac{11}{6}\mathbf{x}_{i-N} + 3\mathbf{x}_{i-N+1} - \tfrac{3}{2}\mathbf{x}_{i-N+2} + \tfrac{1}{3}\mathbf{x}_{i-N+3}\right) +$$
$$\left(\tfrac{N-1}{\tau}\right)^{2}\tfrac{T^2}{2}\left(2\mathbf{x}_{i-N} - 5\mathbf{x}_{i-N+1} + 4\mathbf{x}_{i-N+2} - \mathbf{x}_{i-N+3}\right) + \left(\tfrac{N-1}{\tau}\right)^{3}\tfrac{T^3}{6}\left(-\mathbf{x}_{i-N} + 3\mathbf{x}_{i-N+1} - 3\mathbf{x}_{i-N+2} + \mathbf{x}_{i-N+3}\right) \tag{11}$$

Following the EMHPM procedure, the recursive solution of Equation (1) $\mathbf{X}_{ik}(T)$ is expressed as

$$\mathbf{X}_{ik} = \mathbf{X}_{ik}^{a} + \mathbf{X}_{ik}^{b} + \mathbf{X}_{ik}^{c} + \mathbf{X}_{ik}^{d}, k = 1,2,3\ldots. \tag{12}$$

where

$$\mathbf{X}_{i0}^{a} = \mathbf{x}_{i-1}, \mathbf{X}_{i0}^{b} = \mathbf{X}_{i0}^{c} = \mathbf{X}_{i0}^{d} = 0 \tag{13}$$

and

$$\mathbf{X}_{ik}^{a} = \tfrac{T}{k}\left(\mathbf{A}_t \mathbf{X}_{i(k-1)}^{a} + g(k)\mathbf{B}_t \mathbf{x}_{i-N}\right)$$

$$\mathbf{X}_{ik}^{b} = \tfrac{T}{k+1}\left(\mathbf{A}_t \mathbf{X}_{i(k-1)}^{b} + g(k)\left(\tfrac{N-1}{\tau}\right)\mathbf{B}_t T\left(-\tfrac{11}{6}\mathbf{x}_{i-N} + 3\mathbf{x}_{i-N+1} - \tfrac{3}{2}\mathbf{x}_{i-N+2} + \tfrac{1}{3}\mathbf{x}_{i-N+3}\right)\right)$$

$$\mathbf{X}_{ik}^{c} = \tfrac{T}{k+2}\left(\mathbf{A}_t \mathbf{X}_{i(k-1)}^{c} + g(k)\left(\tfrac{N-1}{\tau}\right)^{2}\mathbf{B}_t \tfrac{T^2}{2}\left(2\mathbf{x}_{i-N} - 5\mathbf{x}_{i-N+1} + 4\mathbf{x}_{i-N+2} - \mathbf{x}_{i-N+3}\right)\right) \tag{14}$$

$$\mathbf{X}_{ik}^{d} = \tfrac{T}{k+3}\left(\mathbf{A}_t \mathbf{X}_{i(k-1)}^{d} + g(k)\left(\tfrac{N-1}{\tau}\right)^{3}\mathbf{B}_t \tfrac{T^3}{6}\left(-\mathbf{x}_{i-N} + 3\mathbf{x}_{i-N+1} - 3\mathbf{x}_{i-N+2} + \mathbf{x}_{i-N+3}\right)\right)$$

The approximate solution of Equation (1) can be obtained by substituting Equation (12) into Equation (10) adding each of the approximations $\mathbf{X}_{ik}$.

2.3. Stability Analysis

To calculate the stability of the differential Equation (1) using the second-order EMHPM, the solution of Equation (10) for second-order EMHPM must be rewritten by grouping each of the discrete values $\mathbf{x}_i$, $\mathbf{x}_{i-N+2}$, $\mathbf{x}_{i-N+1}$, $\mathbf{x}_{i-N}$, resulting in

$$\mathbf{x}_i(T) \approx \mathbf{P}_i(T)\mathbf{x}_{i-1} + \mathbf{Q}_i{}'(T)\mathbf{x}_{i-N+2} + \mathbf{Q}_i(T)\mathbf{x}_{i-N+1} + \mathbf{R}_i(T)\mathbf{x}_{i-N} \tag{15}$$

where

$$\mathbf{P}_i(T) = \sum_{k=0}^{m} \tfrac{1}{k!}\mathbf{A}_t^k T^k,$$

$$\mathbf{Q}'_i(T) = \sum_{k=1}^{m}\left(\tfrac{1}{(k+2)!}\left(\tfrac{N-1}{\tau}\right)^2 \mathbf{A}_t^{k-1}\mathbf{B}_t T^{k+2} - \tfrac{1}{2(k+1)!}\left(\tfrac{N-1}{\tau}\right)\mathbf{A}_t^{k-1}\mathbf{B}_t T^{k+1}\right)$$

$$\mathbf{Q}_i(T) = \sum_{k=1}^{m} \tfrac{1}{(k+1)!}\left(\tfrac{N-1}{\tau}\right)\mathbf{A}_t^{k-1}\mathbf{B}_t T^{k+1} - 2\mathbf{Q}'_i$$

$$\mathbf{R}_i(T) = \sum_{k=1}^{m} \tfrac{1}{k!}\mathbf{A}_t^{k-1}\mathbf{B}_t T^k - \mathbf{Q}'_i - \mathbf{Q}_i$$

$$(16)$$

Similarly, to compute the stability lobes for the third-order EMHPM, the solution of the differential Equation (1) for third-order EMHPM is rewritten as

$$\mathbf{x}_i(T) \approx \mathbf{P}_i(T)\mathbf{x}_{i-1} + \mathbf{Q}''_i(T)\mathbf{x}_{i-N+3} + \mathbf{Q}'_i(T)\mathbf{x}_{i-N+2} + \mathbf{Q}_i(T)\mathbf{x}_{i-N+1} + \mathbf{R}_i(T)\mathbf{x}_{i-N} \tag{17}$$

where

$$\mathbf{P}_i(T) = \sum_{k=0}^{m} \tfrac{1}{k!}\mathbf{A}_t^k T^k,$$

$$\mathbf{Q}''_i(T) = \sum_{k=1}^{m}\left(\tfrac{1}{(k+1)!}\left(\tfrac{N-1}{\tau}\right)\mathbf{A}_t^{k-1}\mathbf{B}_t T^{k+1}\left(\tfrac{1}{3}\right) - \tfrac{1}{(k+2)!}\left(\tfrac{N-1}{\tau}\right)^2 \mathbf{A}_t^{k-1}\mathbf{B}_t T^{k+2} + \tfrac{1}{(k+3)!}\left(\tfrac{N-1}{\tau}\right)^3 \mathbf{A}_t^{k-1}\mathbf{B}_t T^{k+3}\right)$$

$$\mathbf{Q}'_i(T) = \sum_{k=1}^{m}\left[\begin{array}{c}\tfrac{1}{(k+1)!}\left(\tfrac{N-1}{\tau}\right)\mathbf{A}_t^{k-1}\mathbf{B}_t T^{k+1} + \tfrac{1}{(k+2)!}\left(\tfrac{N-1}{\tau}\right)^2 \mathbf{A}_t^{k-1}\mathbf{B}_t T^{k+2}\left(-\tfrac{7}{2}\right) \\[2mm] + \tfrac{1}{(k+3)!}\left(\tfrac{N-1}{\tau}\right)^3 \mathbf{A}_t^{k-1}\mathbf{B}_t T^{k+3}\left(\tfrac{9}{2}\right)\end{array}\right] - \tfrac{15}{2}\mathbf{Q}''_i$$

$$\mathbf{Q}_i(T) = \sum_{k=1}^{m}\left(\tfrac{1}{(k+1)!}\left(\tfrac{N-1}{\tau}\right)\mathbf{A}_t^{k-1}\mathbf{B}_t T^{k+1}\right) - 3\mathbf{Q}''_i - 2\mathbf{Q}'_i$$

$$\mathbf{R}_i(T) = \sum_{k=1}^{m} \tfrac{1}{k!}\mathbf{A}_t^{k-1}\mathbf{B}_t T^k - \mathbf{Q}''_i - \mathbf{Q}'_i - \mathbf{Q}_i$$

$$(18)$$

The approximate solution obtained from Equation (17) was used to define a discrete map following the procedure described in [43]:

$$\mathbf{w}_i = \mathbf{D}_i \mathbf{w}_{i-1} \tag{19}$$

where $\mathbf{w}_{i-1}$ is a vector with dimension equal to the total number of states (displacement and velocity) for all N discrete intervals:

$$\mathbf{w}_{i-1} = \left[\mathbf{x}_{(i-1)}, \dot{\mathbf{x}}_{(i-1)}, \mathbf{x}_{(i-2)}, \ldots, \mathbf{x}_{(i-N)}\right]^T \tag{20}$$

$\mathbf{D}_i$ is a coefficient matrix and for the third-order EMHPM it has the form:

$$\mathbf{D}_i = \begin{bmatrix}
\mathbf{P} & 0 & 0 & 0 & \cdots & 0 & \mathbf{Q}''_i & \mathbf{Q}'_i & \mathbf{Q}_i & \mathbf{R}_i \\
\mathbf{I} & 0 & 0 & 0 & \cdots & 0 & 0 & 0 & 0 & 0 \\
0 & \mathbf{I} & 0 & 0 & \cdots & 0 & 0 & 0 & 0 & 0 \\
0 & 0 & \mathbf{I} & 0 & \cdots & 0 & 0 & 0 & 0 & 0 \\
\vdots & \vdots & \vdots & \ddots & \vdots & \vdots & \vdots & \vdots & \vdots & \vdots \\
0 & 0 & 0 & 0 & \ddots & 0 & 0 & 0 & 0 & 0 \\
0 & 0 & 0 & 0 & \cdots & \mathbf{I} & 0 & 0 & 0 & 0 \\
0 & 0 & 0 & 0 & \cdots & 0 & \mathbf{I} & 0 & 0 & 0 \\
0 & 0 & 0 & 0 & \cdots & 0 & 0 & \mathbf{I} & 0 & 0 \\
0 & 0 & 0 & 0 & \cdots & 0 & 0 & 0 & \mathbf{I} & 0
\end{bmatrix} \tag{21}$$

It is important to point out that in the case of the second-order EMHPM, the matrix $\mathbf{D_i}$ is like the matrix of the third-order EMHPM without the matrix $\mathbf{Q}_i''$.

Then, the Floquet transition matrix $\mathbf{\Phi}$ is calculated over the main period $\tau = (N-1)/\Delta t$, coupling each of the discrete maps $\mathbf{D}_i$, $i - 1, 2, \ldots, (N-1)$, to obtain:

$$\mathbf{\Phi} = \mathbf{D}_{N-1}\mathbf{D}_{N-2}\ldots\mathbf{D}_2\mathbf{D}_1 \tag{22}$$

Thus, the stability of Equation (1) is determined by calculating the eigenvalues of the transition matrix given by Equation (22). The eigenvalues of the transition matrix are actually the Floquet multipliers which are the exponents of each complex exponential functions that describe the motion of Equation (1). If the modulus of greatest magnitude is greater than or equal to one, it implies that the system will behave in an unstable way and the amplitude of the vibration will increase exponentially, otherwise it will have a stable behavior.

3. Numeric Solution of the Milling Equation

3.1. Dynamic Model to the Milling Equation

To validate the proposed EMHPM methods, the numerical solution of the delay differential equation analyzed by Olvera et al., in [33] was calculated, which describes the dynamic model of the milling process in one degree of freedom (DOF):

$$\ddot{\mathbf{x}}(t) + 2\zeta\omega_n\dot{\mathbf{x}}(t) + \omega_n^2\mathbf{x}(t) = -\frac{a_p h_{xx}(t)}{m_m}(\mathbf{x}(t) - \mathbf{x}(t-\tau)) \tag{23}$$

where ζ is the modal damping ratio, ω_n is the natural frequency of the workpiece, a_p is the axial depth of cut, m_m is the modal mass, τ represents the time delay corresponding to the hitting period between each tooth of the tool and $h_{xx}(t)$ is the specific cutting force in the x-direction due to flexibility in x-direction, which was calculated depending on the position of the tool

$$h_{xx}(t) = \sum_{iz=1}^{z_n} g(\phi_{iz}(t))sin\phi_{iz}(t)(K_{tc}\cos\phi_{iz}(t) + K_{nc}\sin\phi_{iz}(t)) \tag{24}$$

z_n is the number of edges of the tool, K_{tc} and K_{nc} are the average specific cut coefficients in the tangential and normal direction, respectively, and $\phi_{iz}(t)$ is the angular position of each left edge described by

$$\phi_{iz} = (2\pi n/60)t + 2\pi iz/z_n \tag{25}$$

where n is the spindle speed in revolution per minute (rpm). The function $g(\phi_{iz}(t))$ is a window function, which has the value of one when the current edge iz is cutting material, otherwise it takes the value zero.

In up-milling $\phi_{st} = 0$ and $\phi_{ex} = \cos^{-1}(1 - 2a_d)$, conversely in down-milling $\phi_{st} = \cos^{-1}(2a_d - 1)$ and $\phi_{ex} = \pi$, a_d is the radial immersion ratio of the cut and ϕ_{st} and ϕ_{ex} are the angular positions where each edge enters and leaves the workpiece.

The second- and third-order EMHPM is applied to obtain the solution of Equation (23) and it is compared with the solution given by the first-order EMHPM [33]; for a regular tool the matrix $\mathbf{A}$ and $\mathbf{B}$ are represented as:

$$\mathbf{A}_t = \begin{bmatrix} 0 & 1 \\ -\omega_n^2 - \frac{a_p h_{xx}(t)}{m_m} & -2\zeta\omega_n \end{bmatrix}, \ \mathbf{B}_t = \begin{bmatrix} 0 & 0 \\ \frac{a_p h_{xx}(t)}{m_m} & 0 \end{bmatrix} \tag{26}$$

$\mathbf{A}_t$ and $\mathbf{B}_t$ correspond to the periodic matrix evaluated at time t. For demonstration purposes, time-domain simulations were computed for a full-immersion down-milling operation. We used the parameters employed by Insperger et al., in [43] where the stability lobes were also calculated.

The modal parameters $f_n = 922$ Hz, $\omega_n = 5793$ rad/s, $\zeta = 0.011$ and $m_m = 0.03993$ kg corresponds to a single degree of freedom. The tangential and normal cutting coefficients are $K_{tc} = 6 \times 10^8$ N/m^2 and $K_{nc} = 2 \times 10^8$ N/m^2 respectively for an end-mill with $z_n = 2$. The time-domain solution was computed using the EMHPM considering $N = 76$ discrete intervals and $m = 7$. Two sets of cutting conditions were chosen for a fixed spindle speed value of $n = 12{,}000$ rpm where the axial depth of cut of $a_p = 1.5$ mm corresponds to a stable cutting operation while that for an unstable operation $a_p = 3$ mm was chosen. In Figure 2 we plot the second- and third-order EMHPM solutions and compare it with the first-order EMHPM and the dde23 routine in Matlab, which is used to integrate DDE.

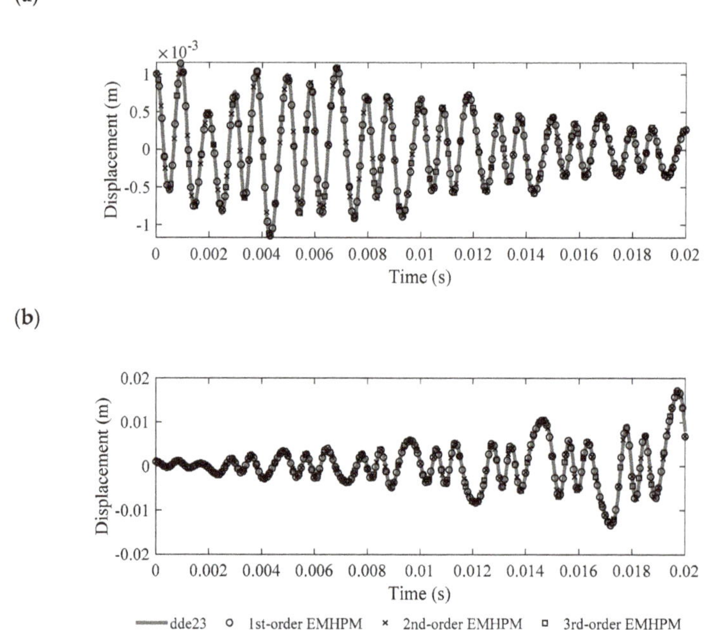

Figure 2. Numerical comparison of the enhanced homotopy perturbation method (EMHPM) solutions of the milling equation, Equation (23), with the dde23 MATLAB routine. (a) Stable milling operation with $a_p = 1.5$ mm, $a_d = 1$ and $n = 12000$ rpm and (b) unstable milling operation with $a_p = 3$ mm, $a_d = 1$ and $n = 12000$ rpm.

3.2. Numerical Comparison between Methods

In order to observe the rate of convergence of the first-, second- and third-order EMHPM, we chose the stable case with cutting conditions $a_p = 1.5$ mm, $a_d = 1$ and $n = 12{,}000$ rpm presented in Figure 3a, and the unstable case with cutting conditions $a_p = 3$ mm, $a_d = 1$ and $n = 12{,}000$ rpm showed in Figure 3b. The rate of convergence was analyzed by computing the absolute error between the solution with N discrete intervals and a converged solution. All methods were compared against itself using the solution provided with $N = 200$ discrete intervals, which are considered the converged solution. In Figure 3a it is observed that the convergence is better for the second- and third-order than the first-order, however, the difference of convergence between second- and third-order with the parameters used was negligible. On the other hand, Figure 3b shows that for few discrete intervals the third-order EMHPM had the fastest convergence in comparison with the second- and the first-order EMHPM. However, the second-order and third-order curves behaved very similarly after $N = 50$ discrete intervals. It is important to mention that for a typical stability solution in the ranges of spindle speed 5000–10,000 rpm, $N = 40$ discrete intervals will be enough to have accurate predictions.

(**a**)

(**b**)

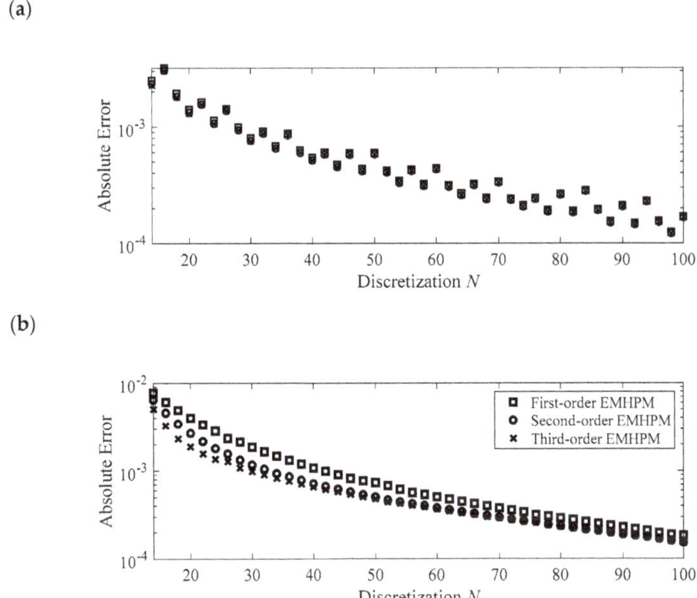

Figure 3. Convergence rate of absolute error between first-, second- and third-order EMHPM for down-milling operation. Cutting parameters for (**a**) $a_p = 1.5$ mm, $a_d = 1$ and $n = 12,000$ rpm and (**b**) $a_p = 3$ mm, $a_d = 1$ and $n = 12,000$ rpm.

Since the rate of convergence was proved for time-domain simulations, we next explored the convergence of the methods applied to the stability analysis. The stability lobes computed with the second- and third-order EMHPM for regular milling tools were compared with its predecessor for radial immersion value of $a_d = 1$ and the other parameters indicated above as it was used in [44]. Figure 4 shows the stability diagrams for spindle speed in the range 2000–3000 rev/min where the precision of the method was compromised due to the higher value of the time delay. While the shaded gray area represents the stability lobes computed with $N = 200$ discrete intervals in all subfigures, in each subfigure solid black lines draw the stability frontier for a specific discrete interval and using the first-, second- or third-order EMHPM. In Figure 4 the first, second and third column represents the solution for the first-, the second- and the third-order EMHPM respectively, while the first and the second row was for $N = 60$ and $N = 100$ discrete intervals, respectively. It is observed that the error achieved in the third-order EMHPM was less than those attained for the first-order and second-order EMHPM solutions. This confirms that the third-order EMHPM had the highest rate of convergence.

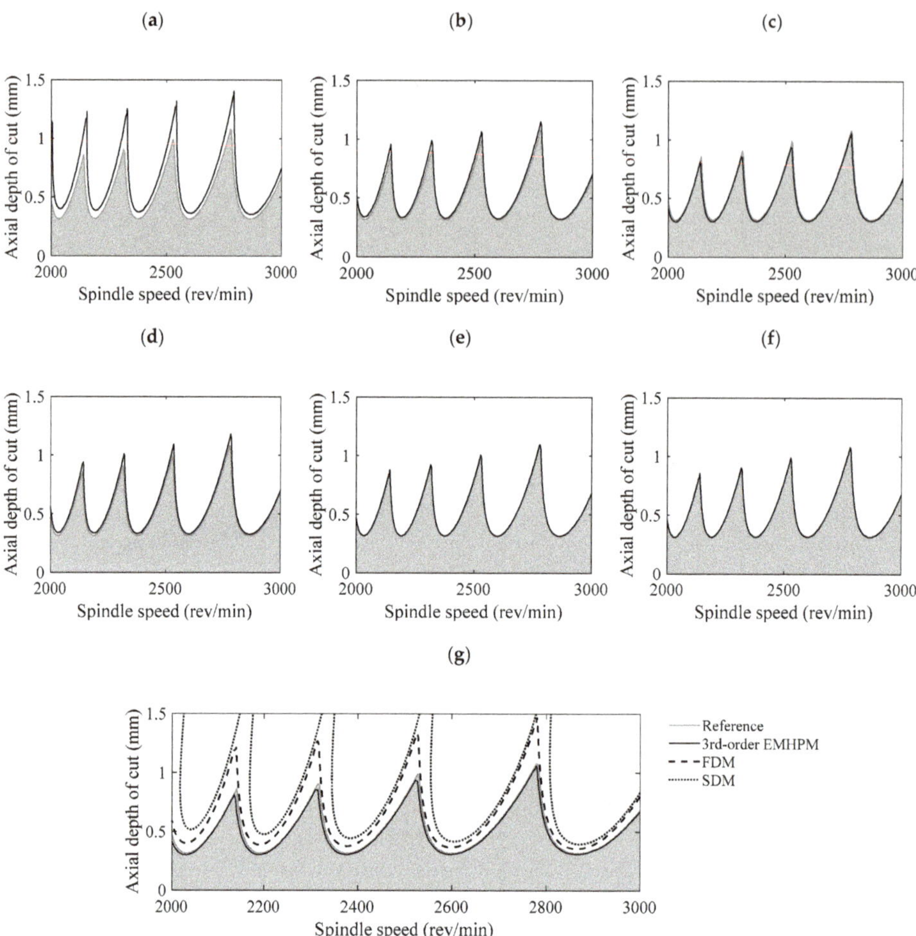

Figure 4. Stability diagrams for down-milling operation. The first (**a–c**) and second (**d–f**) rows of subfigures correspond to N = 60 and N = 100 discrete intervals, respectively. First (**a,d**), second (**b,e**) and third (**c,f**) columns correspond to the first-, second- and third-order EMHPM, respectively. Subfigure (**g**) shows a comparison between the third-order EMHPM (black line), the SDM (dot line) and the FDM (dash line) with N = 60 discrete intervals.

The results were also compared, in Figure 4g, with the semi-discretization method (SDM) presented by Insperger and Stépán in [43] (dot line) and with the full-discretization method (FDM) presented by Ding et al. in [45] (dash line). It is observed that the EMHPM converged faster than the SDM and FDM. Table 1 list the results for computation times for a different number of discretized intervals together with the absolute error between stability frontiers. Notice that the solution obtained with the EMHPM with $N = 60$ discrete intervals was faster than the SDM and the FDM and even the error was less in the solution by the EMHPM. For $N = 100$ discrete intervals the computation time for the FDM was similar to the solution obtained with the EMHPM but it was demonstrated that the solution by second- and third-order EMHPM requires a smaller number of discretized intervals to converge to the solution and using a smaller amount of computation time. Notice the error for the SDM was not calculated since there was no stability frontier in some values of spindle speed.

Table 1. Comparison of convergence for different methods for down-milling operation with $a_d = 1$

Method	N = 60		N = 100	
	Time (s)	Error	Time (s)	Error
First-order EMHPM	52.39	1.6276	186.35	0.3203
Second-order EMHPM	54.85	0.3507	194.65	0.0568
Third-order EMHPM	61.48	0.3458	195.86	0.0563
FDM	61.26	1.4097	191.49	0.3998
SDM	278.03	-	570.04	-

It is noticeable that exists a significant improvement in the rate of convergence from first-order to second-order and third-order EMHPM, however, the difference between the second- and the third-order EMHPM is negligible if the number of discretized intervals increase. There is no best method between second- and third-order EMHPM in terms of rate of convergence and computation time since the precision depends on the nature of the studied problem. However, it is easy to prove that a higher-order approximation (fourth- and fifth-order EMHPM) could drastically increase the computational time without a significant improvement in the solution.

4. Stability Analysis of Multivariable Milling Tools

The EMHPM can be generalized for stability analysis of DDEs having multiple delays. A multivariable tool contains some of the following characteristics: uneven pitch between teeth, and/or at least one helix angle with a different value from the others. This analysis was developed by Compeán et al., in [34] by using the first-order EMHPM, where the methodology for the characterization of the cutting coefficients for a multivariable tool was discussed, and the dynamic behavior was studied from the productivity point of view. Since the angular spacing at the beginning of the edge is different between teeth (pitch) and the different values of helix angles of the edges between adjacent teeth, the angular spacing between teeth at a specific height changes continuously, which produces an infinite number of delays. A common approach to deal with the DDE with an infinite number of delay is to discretize the tool by cutting disks in the axial direction with a thickness Δa_{dsk} to induce a DDE with a finite number of delays. A single disk still has the same number of flutes (discrete flutes) and considering that the maximum delay in the process is the period of rotation of the tool or the spindle rotation period τ_T, then, it can be discretized in $N-1$ intervals.

The angular position between two adjacent teeth in each cutting disk changes according to the axial position of the referred disk and is related to the expression $\psi = k_\beta a_p$, where $k_\beta = 2\tan\beta/2D$. Here D is the diameter of the tool and ψ represents the cutting-edge offset angle due to the helix angle. A certain interval can be associated with a discrete time delay of each tooth iz and disk l using the following formulation

$$N_{iz,l} = round\left((N-1)\frac{\delta\phi_{iz,l}}{2\pi}\right) \tag{27}$$

where $\delta\phi_{iz,l}$ is the angular pitch between consecutive teeth for each disk, the round function converts the argument to the nearest integer. In Equation (27) $N_{iz,}$ is a table (matrix) of dimension $iz \times l$. Since this procedure could generate several delayed terms and some of them with the same value of discrete time delay due to the discretization scheme, it is required to collect all the different (non-repeated) discrete time delays d_n from $N_{iz,l}$.

Thus, without loss of generality, the DDE with multiple delays can be written as

$$\dot{x}(t) = A(t)x(t) + \sum_{d=\min(d_n)}^{\max(d_n)} B^d x(t-\tau) \tag{28}$$

where $\mathbf{x}$ is the vector of states, $\mathbf{A}(t + \tau_T) = \mathbf{A}(\tau_T)$, $\mathbf{B}^d(t + \tau_T) = \mathbf{B}^d(\tau_T)$ and τ_T is the period of rotation of the spindle. Following the EMHPM procedure, Equation (28) can be written equivalently by intervals as:

$$\dot{\mathbf{x}}_i(T) - \mathbf{A}_t\mathbf{x}_i(T) \approx \sum_{d=\min(d_n)}^{\max(d_n)} \mathbf{B}_t{}^d\mathbf{x}_i{}^{\tau_d}(T) \tag{29}$$

being $\mathbf{x}_i(T)$ the solution by intervals of order m for Equation (28) in the $i-th$ interval that satisfies the initial condition $\mathbf{x}_i(0) = \mathbf{x}_{i-1}$, the matrices $\mathbf{A}_t$ and $\mathbf{B}_t^d$ represent the values of the matrices $\mathbf{A}(t)$ and $\mathbf{B}^d(t)$ evaluated at time t respectively.

4.1. Third-Order EMHPM for Multivariable Milling Tool

To approximate the term associated with the delayed terms $\mathbf{x}_i^{\tau_d}(T)$ of Equation (29), the interval of the period τ_T, $[t_0 - \tau_T,\ t_0]$ is discretized in $N - 1$ intervals that can be equal size. For simplicity, intervals of equal size $\Delta t = \tau_T/(N-1)$ are chosen. Then it is assumed that the function $\mathbf{x}_i^{\tau_d}(T)$, which is defined in the interval $[t_{i-d-1},\ t_{i-d+2}]$, for the third-order EMHPM has the representation of the form:

$$\mathbf{x}_i^{\tau_d}(T) = \mathbf{x}_{i-d}(T) \approx \mathbf{x}_{i-d-1} + \left(\tfrac{N-1}{\tau_T}\right)T\left(-\tfrac{11}{6}(\mathbf{x}_{i-d-1}) + 3(\mathbf{x}_{i-d}) - \tfrac{3}{2}(\mathbf{x}_{i-d+1}) + \tfrac{1}{3}(\mathbf{x}_{i-d+2})\right) +$$
$$\left(\tfrac{N-1}{\tau_T}\right)^2\tfrac{T^2}{2}\left(2(\mathbf{x}_{i-d-1}) - 5(\mathbf{x}_{i-d}) + 4(\mathbf{x}_{i-d+1}) - (\mathbf{x}_{i-d+2})\right) + \tag{30}$$
$$\left(\tfrac{N-1}{\tau_T}\right)^3\tfrac{T^3}{6}\left(-\mathbf{x}_{i-d-1} + 3(\mathbf{x}_{i-d}) - 3(\mathbf{x}_{i-d+1}) + \mathbf{x}_{i-d+2}\right)$$

Defining $\mathbf{x}_i \equiv \mathbf{x}_i(T_i)$ to simplify the notation, and substituting Equation (30) in Equation (29), the following equation is obtained:

$$\dot{\mathbf{x}}_i(T) = \mathbf{A}_t\mathbf{x}_i(T) + \sum_{d=\min(d_n)}^{\max(d_n)} \left\{ \begin{array}{l} \mathbf{B}_t{}^d\mathbf{x}_{i-d-1} + \mathbf{B}_t{}^dT\left(\tfrac{N-1}{\tau_T}\right)\left(-\tfrac{11}{6}(\mathbf{x}_{i-d-1}) + 3(\mathbf{x}_{i-d}) - \tfrac{3}{2}(\mathbf{x}_{i-d+1}) + \tfrac{1}{3}(\mathbf{x}_{i-d+2})\right) + \\[4pt] \mathbf{B}_t{}^d\tfrac{T^2}{2}\left(\tfrac{N-1}{\tau_T}\right)^2\left(2(\mathbf{x}_{i-d-1}) - 5(\mathbf{x}_{i-d}) + 4(\mathbf{x}_{i-d+1}) - \mathbf{x}_{i-d+2}\right) + \\[4pt] \mathbf{B}_t{}^d\tfrac{T^3}{6}\left(\tfrac{N-1}{\tau_T}\right)^3\left(-\mathbf{x}_{i-d-1} + 3(\mathbf{x}_{i-d}) - 3(\mathbf{x}_{i-d+1}) + \mathbf{x}_{i-d+2}\right) \end{array} \right\} \tag{31}$$

where

$$\mathbf{A}_t = \begin{bmatrix} 0 & 1 \\ -\omega_n^2 - \tfrac{\Delta a_{disk}}{m_m}\sum_{d=\min(d_n)}^{\max(d_n)} h_{yy}^d & -2\zeta\omega_n \end{bmatrix}, \quad \mathbf{B}_t{}^d = \begin{bmatrix} 0 & 0 \\ \tfrac{\Delta a_{disk}}{m_m}\sum_{d=\min(d_n)}^{\max(d_n)} h_{yy}^d & 0 \end{bmatrix} \tag{32}$$

here, h_{yy} is the specific cutting force in the y-direction due to flexibility in y-direction, which is used for thin wall machining. This force was calculated depending on the position of the tool via the following equation:

$$h_{yy}(t) = \sum_{iz=1}^{z_n} g(\phi_{iz}(t)) \cos\phi_{iz}(t)\left(-K_{tc}\sin\phi_{iz}(t) + K_{nc}\cos\phi_{iz}(t)\right) \tag{33}$$

Then, solving Equation (31) yields

$$\mathbf{X}_{i0} = \mathbf{x}_{i-1}$$

$$\mathbf{X}_{i1} = T\mathbf{A}_t(\mathbf{x}_{i-1}) + \sum_{d=\min(d_n)}^{\max(d_n)} \left(\begin{array}{l} TD_t{}^d(\mathbf{x}_{i-d-1}) + B_t{}^d \frac{T^2}{2}\left(\frac{N-1}{\tau_T}\right)\left(-\frac{11}{6}(\mathbf{x}_{i-d-1}) + 3(\mathbf{x}_{i-d}) - \frac{3}{2}(\mathbf{x}_{i-d+1}) + \frac{1}{3}(\mathbf{x}_{i-d+2})\right) \\[2mm] + B_t{}^d \frac{T^3}{6}\left(\frac{N-1}{\tau_T}\right)^2 (2(\mathbf{x}_{i-d-1}) - 5(\mathbf{x}_{i-d}) + 4(\mathbf{x}_{i-d+1}) - \mathbf{x}_{i-d+2}) \\[2mm] + B_t{}^d \frac{T^4}{24}\left(\frac{N-1}{\tau_T}\right)^3 (-\mathbf{x}_{i-d-1} + 3(\mathbf{x}_{i-d}) - 3(\mathbf{x}_{i-d+1}) + \mathbf{x}_{i-d+2}) \end{array} \right)$$

$$\mathbf{X}_{i2} = \frac{T^2}{2}\mathbf{A}_t{}^2(\mathbf{x}_{i-1}) + \mathbf{A}_t \sum_{d=\min(d_n)}^{\max(d_n)} \left(\begin{array}{l} \frac{T^2}{2}B_t{}^d(\mathbf{x}_{i-d-1}) + B_t{}^d \frac{T^3}{6}\left(\frac{N-1}{\tau_T}\right)\left(-\frac{11}{6}(\mathbf{x}_{i-d-1}) + 3(\mathbf{x}_{i-d}) - \frac{3}{2}(\mathbf{x}_{i-d+1}) + \frac{1}{3}(\mathbf{x}_{i-d+2})\right)+ \\[2mm] B_t{}^d \frac{T^4}{24}\left(\frac{N-1}{\tau_T}\right)^2 (2(\mathbf{x}_{i-d-1}) - 5(\mathbf{x}_{i-d}) + 4(\mathbf{x}_{i-d+1}) - \mathbf{x}_{i-d+2})+ \\[2mm] B_t{}^d \frac{T^5}{120}\left(\frac{N-1}{\tau_T}\right)^3 (-\mathbf{x}_{i-d-1} + 3(\mathbf{x}_{i-d}) - 3(\mathbf{x}_{i-d+1}) + \mathbf{x}_{i-d+2}) \end{array} \right) \tag{34}$$

$$\vdots$$

$$\mathbf{X}_{ik} = \frac{1}{k!}T^k\mathbf{A}_t{}^k(\mathbf{x}_{i-1}) + \mathbf{A}_t{}^{k-1} \sum_{d=\min(d_n)}^{\max(d_n)} \left(\begin{array}{l} \frac{1}{k!}T^k B_t{}^d(\mathbf{x}_{i-d-1}) + \frac{1}{(k+1)!}T^{k+1} B_t{}^d\left(\frac{N-1}{\tau_T}\right)\left(-\frac{11}{6}(\mathbf{x}_{i-d-1}) + 3(\mathbf{x}_{i-d}) - \frac{3}{2}(\mathbf{x}_{i-d+1})+ \right. \\[2mm] \frac{1}{3}(\mathbf{x}_{i-d+2})\Big) + \frac{1}{(k+2)!}T^{k+2} B_t{}^d\left(\frac{N-1}{\tau_T}\right)^2 (2(\mathbf{x}_{i-d-1}) - 5(\mathbf{x}_{i-d}) + 4(\mathbf{x}_{i-d+1}) - \mathbf{x}_{i-d+2})+ \\[2mm] \frac{1}{(k+3)!}T^{k+3} B_t{}^d\left(\frac{N-1}{\tau_T}\right)^3 (-\mathbf{x}_{i-d-1} + 3(\mathbf{x}_{i-d}) - 3(\mathbf{x}_{i-d+1}) + \mathbf{x}_{i-d+2}) \end{array} \right)$$

Notice that Equation (34) can be written recursively as

$$\mathbf{X}_{ik} = \mathbf{X}_{ik}^a + \mathbf{X}_{ik}^b + \mathbf{X}_{ik}^c + \mathbf{X}_{ik}^d, k = 1, 2, 3 \ldots. \tag{35}$$

where $\mathbf{X}_{i0}^a = \mathbf{x}_{i-1}$, $\mathbf{X}_{i0}^b = \mathbf{X}_{i0}^c = \mathbf{X}_{i0}^d = 0$ and

$$\mathbf{X}_{ik}^a = \frac{T}{k}\left(\mathbf{A}_t \mathbf{X}_{i(k-1)}^a + g(k) \sum_{d=\min(d_n)}^{\max(d_n)} \mathbf{B}_t^d \mathbf{x}_{i-d-1} \right)$$

$$\mathbf{X}_{ik}^b = \frac{T}{k+1}\left(\mathbf{A}_t \mathbf{X}_{i(k-1)}^b + g(k) \sum_{d=\min(d_n)}^{\max(d_n)} \left(\frac{N-1}{\tau}\right)\mathbf{B}_t^d T\left(-\frac{11}{6}(\mathbf{x}_{i-d-1}) + 3(\mathbf{x}_{i-d}) - \frac{3}{2}(\mathbf{x}_{i-d+1}) + \frac{1}{3}(\mathbf{x}_{i-d+2})\right) \right)$$

$$\mathbf{X}_{ik}^c = \frac{T}{k+2}\left(\mathbf{A}_t \mathbf{X}_{i(k-1)}^c + g(k) \sum_{d=\min(d_n)}^{\max(d_n)} \left(\frac{N-1}{\tau}\right)^2 \mathbf{B}_t^d \frac{T^2}{2}(2(\mathbf{x}_{i-d-1}) - 5(\mathbf{x}_{i-d}) + 4(\mathbf{x}_{i-d+1}) - \mathbf{x}_{i-d+2}) \right)$$

$$\mathbf{X}_{ik}^d = \frac{T}{k+3}\left(\mathbf{A}_t \mathbf{X}_{i(k-1)}^d + g(k) \sum_{d=\min(d_n)}^{\max(d_n)} \left(\frac{N-1}{\tau}\right)^3 \mathbf{B}_t^d \frac{T^3}{6}(-\mathbf{x}_{i-d-1} + 3(\mathbf{x}_{i-d}) - 3(\mathbf{x}_{i-d+1}) + \mathbf{x}_{i-d+2}) \right) \tag{36}$$

the solution of order m for Equation (31) was obtained by adding each of the approximations k of Equation (35). Similar to Equation (15), to obtain the stability graphs the solution of Equation (35) is rewritten by grouping the discrete states, which results in:

$$\mathbf{x}_i(T) \approx \mathbf{P}_i(T)\mathbf{x}_{(i-1)} + \sum_{d=\min(d_n)}^{\max(d_n)} \left(\mathbf{Q}_i^{\prime\prime d}(T)\mathbf{x}_{i-d+2} + \mathbf{Q}_i^{\prime d}(T)\mathbf{x}_{i-d+1} + \mathbf{Q}_i^d(T)\mathbf{x}_{i-d} + \mathbf{R}_i^d(T)\mathbf{x}_{i-d-1} \right) \tag{37}$$

where

$$\mathbf{P}_i(T) = \sum_{k=0}^{m} \frac{1}{k!} \mathbf{A}_t^k T^k,$$

$$\mathbf{Q}_i^{''d}(T) = \sum_{k=1}^{m} \left(\begin{array}{l} \frac{1}{(k+1)!}\left(\frac{N-1}{\tau_T}\right)\mathbf{A}_t^{k-1}\mathbf{B}_t^d T^{k+1}\left(\frac{1}{3}\right) - \frac{1}{(k+2)!}\left(\frac{N-1}{\tau_T}\right)^2 \mathbf{A}_t^{k-1}\mathbf{B}_t^d T^{k+2} + \\[2mm] \frac{1}{(k+3)!}\left(\frac{N-1}{\tau_T}\right)^3 \mathbf{A}_t^{k-1}\mathbf{B}_t^d T^{k+3} \end{array} \right)$$

$$\mathbf{Q}_i^{'d}(T) = \sum_{k=1}^{m} \left(\begin{array}{l} \frac{1}{(k+1)!}\left(\frac{N-1}{\tau_T}\right)\mathbf{A}_t^{k-1}\mathbf{B}_t^d T^{k+1} + \frac{1}{(k+2)!}\left(\frac{N-1}{\tau_T}\right)^2 \mathbf{A}_t^{k-1}\mathbf{B}_t^d T^{k+2}\left(-\frac{7}{2}\right) + \\[2mm] \frac{1}{(k+3)!}\left(\frac{N-1}{\tau_T}\right)^3 \mathbf{A}_t^{k-1}\mathbf{B}_t^d T^{k+3}\left(\frac{9}{2}\right) \end{array} \right) - \frac{15}{2}\mathbf{Q}_i^{''d} \qquad (38)$$

$$\mathbf{Q}_i^{d}(T) = \sum_{k=1}^{m} \left(\frac{1}{(k+1)!}\left(\frac{N-1}{\tau_T}\right)\mathbf{A}_t^{k-1}\mathbf{B}_t^d T^{k+1} \right) - 3\mathbf{Q}_i^{''d} - 2\mathbf{Q}_i^{'d}$$

$$\mathbf{R}_i^{d}(T) = \sum_{k=1}^{m} \frac{1}{k!}\mathbf{A}_t^{k-1}\mathbf{B}_t^d T^{k} - \mathbf{Q}_i^{''d} - \mathbf{Q}_i^{'d} - \mathbf{Q}_i^{d}$$

The approximate solution obtained from Equation (37) was used to define a discrete map:

$$\mathbf{w}_i = \mathbf{D}_i \mathbf{w}_{i-1} \qquad (39)$$

where $\mathbf{w}_{i-1}$ is a vector represented by Equation (20) and $\mathbf{D}_i$ is a coefficient matrix given by

$$\mathbf{D}_i = \begin{bmatrix}
\mathbf{P}_i & 0 & \cdots & \mathbf{Q}_i^{''1} & \mathbf{Q}_i^{'1} & \mathbf{Q}_i^{1} & \mathbf{R}_i^{1} & \cdots & \mathbf{Q}_i^{''2} & \mathbf{Q}_i^{'2} & \mathbf{Q}_i^{2} & \mathbf{R}_i^{2} & \cdots & \mathbf{Q}_i^{''d_n} & \mathbf{Q}_i^{'d_n} & \mathbf{Q}_i^{d_n} & \mathbf{R}_i^{d_n} & \cdots & 0 \\
\mathbf{I} & 0 & \cdots & 0 & 0 & 0 & 0 & \cdots & 0 & 0 & 0 & 0 & \cdots & 0 & 0 & 0 & 0 & \cdots & 0 \\
0 & \mathbf{I} & \cdots & 0 & 0 & 0 & 0 & \cdots & 0 & 0 & 0 & 0 & \cdots & 0 & 0 & 0 & 0 & \cdots & 0 \\
0 & 0 & \ddots & 0 & 0 & 0 & 0 & \cdots & 0 & 0 & 0 & 0 & \cdots & 0 & 0 & 0 & 0 & \cdots & 0 \\
\vdots & \vdots & \vdots & \mathbf{I} & 0 & 0 & 0 & \cdots & 0 & 0 & 0 & 0 & \cdots & 0 & 0 & 0 & 0 & \cdots & 0 \\
0 & 0 & \cdots & 0 & \mathbf{I} & 0 & 0 & \cdots & 0 & 0 & 0 & 0 & \cdots & 0 & 0 & 0 & 0 & \cdots & 0 \\
0 & 0 & \cdots & 0 & 0 & \mathbf{I} & 0 & \cdots & 0 & 0 & 0 & 0 & \cdots & 0 & 0 & 0 & 0 & \cdots & 0 \\
0 & 0 & \cdots & 0 & 0 & 0 & \mathbf{I} & \cdots & 0 & 0 & 0 & 0 & \cdots & 0 & 0 & 0 & 0 & \cdots & 0 \\
0 & 0 & \cdots & 0 & 0 & 0 & 0 & \ddots & 0 & 0 & 0 & 0 & \cdots & 0 & 0 & 0 & 0 & \cdots & 0 \\
0 & 0 & \cdots & 0 & 0 & 0 & 0 & \cdots & \mathbf{I} & 0 & 0 & 0 & \cdots & 0 & 0 & 0 & 0 & \cdots & 0 \\
0 & 0 & \cdots & 0 & 0 & 0 & 0 & \cdots & 0 & \mathbf{I} & 0 & 0 & \cdots & 0 & 0 & 0 & 0 & \cdots & 0 \\
0 & 0 & \cdots & 0 & 0 & 0 & 0 & \cdots & 0 & 0 & \mathbf{I} & 0 & \cdots & 0 & 0 & 0 & 0 & \cdots & 0 \\
0 & 0 & \cdots & 0 & 0 & 0 & 0 & \cdots & 0 & 0 & 0 & \mathbf{I} & \cdots & 0 & 0 & 0 & 0 & \cdots & 0 \\
0 & 0 & \cdots & 0 & 0 & 0 & 0 & \cdots & 0 & 0 & 0 & 0 & \ddots & 0 & 0 & 0 & 0 & \cdots & 0 \\
0 & 0 & \cdots & 0 & 0 & 0 & 0 & \cdots & 0 & 0 & 0 & 0 & \cdots & \mathbf{I} & 0 & 0 & 0 & \cdots & 0 \\
0 & 0 & \cdots & 0 & 0 & 0 & 0 & \cdots & 0 & 0 & 0 & 0 & \cdots & 0 & \mathbf{I} & 0 & 0 & \cdots & 0 \\
0 & 0 & \cdots & 0 & 0 & 0 & 0 & \cdots & 0 & 0 & 0 & 0 & \cdots & 0 & 0 & \mathbf{I} & 0 & \cdots & 0 \\
\vdots & \vdots & \cdots & \vdots & \vdots & \vdots & \vdots & \cdots & \vdots & \vdots & \vdots & \vdots & \cdots & 0 & 0 & 0 & \mathbf{I} & \cdots & \vdots \\
0 & 0 & \cdots & 0 & 0 & 0 & 0 & \cdots & 0 & 0 & 0 & 0 & \cdots & 0 & 0 & 0 & 0 & \cdots & 0
\end{bmatrix} \qquad (40)$$

The transition matrix $\boldsymbol{\Phi}$ over the period $\tau_T = (N-1)/\Delta t$ was determined by coupling each solution $\mathbf{x}_i$ through the discrete map D_i, $i = 1, 2, \ldots, (N-1)$. However, the computational cost can be reduced by computing only the transition matrix up to the maximum delayed term without losing precision in the calculation of the eigenvalues:

$$\boldsymbol{\Phi} = \mathbf{D}_{Nmax}\mathbf{D}_{Nmax-1}\cdots\mathbf{D}_2\mathbf{D}_1 \qquad (41)$$

Thus, the stability graphs of Equation (28) were determined by computing the eigenvalues of the transition matrix of Equation (41). The results obtained from the EMHPM were corroborated with the stability lobes in the study of multivariate tools [27].

4.2. Experimental Characterization of One Degree of Freedom Milling Equation and Cutting Force Model

4.2.1. Experimental Modal Analysis

An experimental workpiece was assembled with a 7075T6 aluminum block of 101 mm × 172 mm supported by two thin plates (walls) with a thickness of 4.5 mm. This assembly mimics a DOF as described in Equation (23). The workpiece assembly was rigidly fixed to the workbench of a Makino F3 machining center. For modal analysis, tap testing was performed using a 352C68 PCB Piezotronics accelerometer and an impact hammer model 9722A500. The signals were acquired with a Polytec VIB-E-220 data acquisition card and processed with VibSoft signal analyzer software as shown in Figure 5a. Using the CutPro 8 software, the modal parameters were fitted resulting the values $\zeta = 0.068$, $m_m = 3.8$ kg, $f_m = 132$ Hz and $\omega_n = 829$ rad/s.

(a)   **(b)**

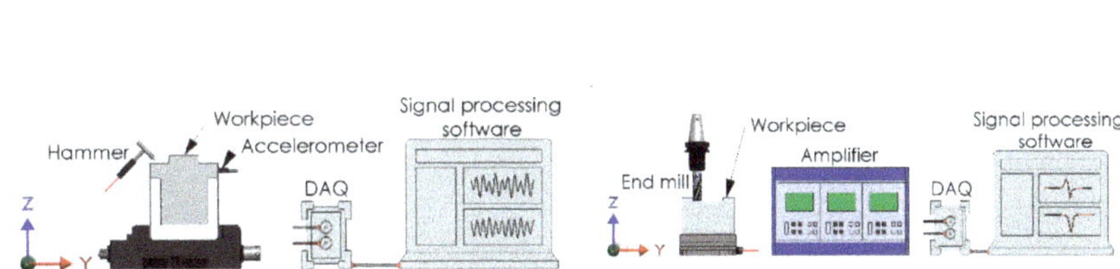

Figure 5. Scheme of the experimental setup for (**a**) the modal analysis and (**b**) cutting forces characterization.

4.2.2. Experimental Determination of Cutting Coefficients

The force model in Equation (33) was used to predict the cutting force magnitude for a given depth of cut. It is based on a mechanistic approach that assumes a relationship between forces and the uncut chip thickness by means of the cutting coefficients. The cutting force model was established by introducing cutting (shearing) and edge coefficients for the tangential and normal directions of the milling tool. The characterization procedure assumed the linear relationship between the averaged experimental cutting forces $\widetilde{F}$ and the feed rate f_z in x- and y- directions. This relationship is established as follows:

$$\widetilde{F} = f_z \widetilde{F}_c + \widetilde{F}_e \tag{42}$$

Here, $\widetilde{F}_c$ and $\widetilde{F}_e$ are the cutting shear and edge components, respectively. The experimental forces at each feed rate are measured, and the cutting-edge components $\widetilde{F}_c$ and $\widetilde{F}_e$ were evaluated

$$K_{tc} = 4\frac{\widetilde{F}_{yc}}{z_n a_p}, \; K_{nc} = -4\frac{\widetilde{F}_{xc}}{z_n a_p} \tag{43}$$

A multivariable cutter provided by a local toolmaker was characterized by using Equation (43) and the experimental setup shown in Figure 5b. Table 2 summarizes the main geometric characteristics of the multivariable tool. A total of five cuttings were performed for full radial immersion in aluminum 7075T6 during dry machining. The forces were recorded by using a dynamometer 9257B Kistler and the spindle speed was set at 3000 rpm based on the dynamometer's natural frequency to avoid the amplification of milling forces. The force signals were acquired using a VibSoft-20 acquisition card at a sample rate of 48 kHz and processed in a custom-made MATLAB app to remove drift and noise. Cutting forces data were collected for the axial depth of cut of 2 mm and four values of feed per tooth 0.05, 0.10, 0.015 and 0.20, so the resulting cutting coefficients K_{tc} for the tooth 1, 2, 3 and 4 were 1215×10^6, 1369×10^6, 897×10^6 and 1799×10^6 N/m² respectively, while that the coefficients K_{nc} for the tooth 1, 2, 3 and 4 resulted 272×10^6, 520×10^6, 801×10^6 and 859×10^6 N/m² respectively.

Table 2. Main geometric parameters of multivariable tool.

	Diameter	12.7 mm
	Cutting length	25 mm
	Coating type	Uncoated
	Number of teeth	4
	Helix angles	39°, 37°, 39°, 41°
	Pitch angles	80°, 100°, 70°, 110°

4.3. Stability Analysis of 1 DOF Milling with a Multivariable Tool

The stability lobes computed for the multivariable tool using the third-order EMHPM with a mesh of 400×200 ($n \times a_p$) are shown in Figure 6 together with stability lobes for a regular tool (angles of 90° and helix angles of 30° for all flutes). An approximation of order $m = 7$ was used with $N = 241$ and $a_d = 1$ mm. Notice from Figure 6 that the stable zone obtained for the multivariable tool was significantly larger, meaning that the critical depth of cut was higher in most spindle speeds, which allowed having more global productivity. It is also observed in the range of spindle speed between 2000 and 3000 rpm, a stable peninsula formed with axial depth ranging from 11 to 20 mm or higher values of critical depth of cut. For instance, for the multivariable cutter at 2500 rpm, the critical depth of cut a_p was 2.17 mm, however it became stable again as shown in Figure 6 for the interval values between 11 and 20 mm. To validate this unexpected behavior, we performed several time-domain simulations using the third-order EMHPM solution described by Equation (37).

(a)   (b)

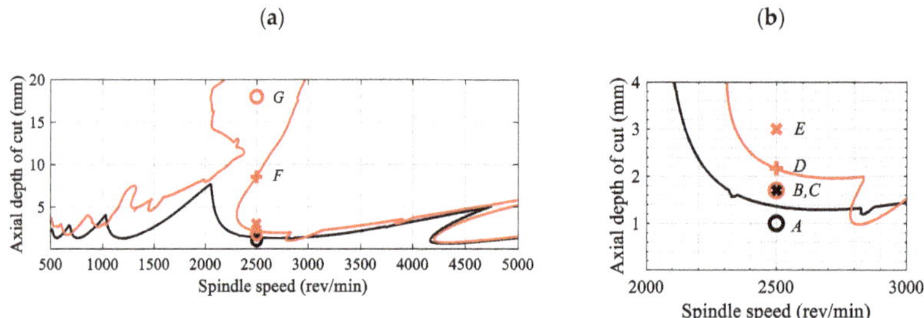

Figure 6. (**a**) Comparison of stability lobes for regular (black solid line) and multivariable (red solid line) cutters by using the third-order EMHPM and (**b**) zoom in on chosen cutting conditions for time-domain simulations. The selected points are marked as follows: unstable (cross mark), stable (circle mark) and transition (plus mark) cutting conditions.

Furthermore, the simulated vibrations for the chosen cutting conditions were analyzed using the continuous wavelet transform (CWT), the power spectral density (PSD) and Poincaré maps (PM). The CWT is a time-frequency representation of a signal that offers the capability to observe how frequencies evolve in time. The scalograms display the absolute value of CWT of the simulated vibration and therefore, they were used to detect chatter phenomena that appeared when milling with a multivariable tool. The PSD is based on the Fourier transform that provides the transformation from the time-domain to the frequency-domain. Additionally, PSD is defined as the squared value of the signal and describes the power of a signal or time series distributed over different frequencies [46]. Moreover, a PM represents points in phase space, which are sampled every spindle rotation [47].

The frequencies f of the CWT and PM were normalized $f_n = f / f_h$ according to the spindle frequency f_h. When milling with a regular milling tool the excitation frequency f_e is equal to z_n times frequencies of the spindle speed f_h but in a multivariable tool, there are several excitation frequencies since the angular spacing between teeth change as a function of the axial depth of cut.

Figure 7 illustrates the CWT, PSD and PM for simulated vibrations using the multivariable tool with different axial depths denoted as cutting conditions A, B and C for the axial depths of cut of 1.0, 1.7 and 1.7 mm respectively. Figure 7a–c refers to the vibrations of the cutting conditions A marked in Figure 6, using a regular tool. The scalogram in Figure 7a identifies point A as a stable cutting since normalized cutting frequencies present a dominant value of $f_n = 3.2$, which corresponds to the natural frequency $f_m = 132$ Hz. This is also confirmed by the PSD analysis shown in Figure 7b. The PM illustrated in Figure 7c shows a vibration that decreased with time and sampled data concentrated in the center confirmed a typical stable case. When the axial depth of cut was increased to 1.7 mm, the stability diagram predicted unstable cutting conditions according to the stability lobes for the regular tool. This case is denoted with cutting conditions B and the corresponding scalogram (shown in Figure 7d) illustrated how the intensity of the dominant frequency increased with time even when the excitation frequency was the same as the case in A.

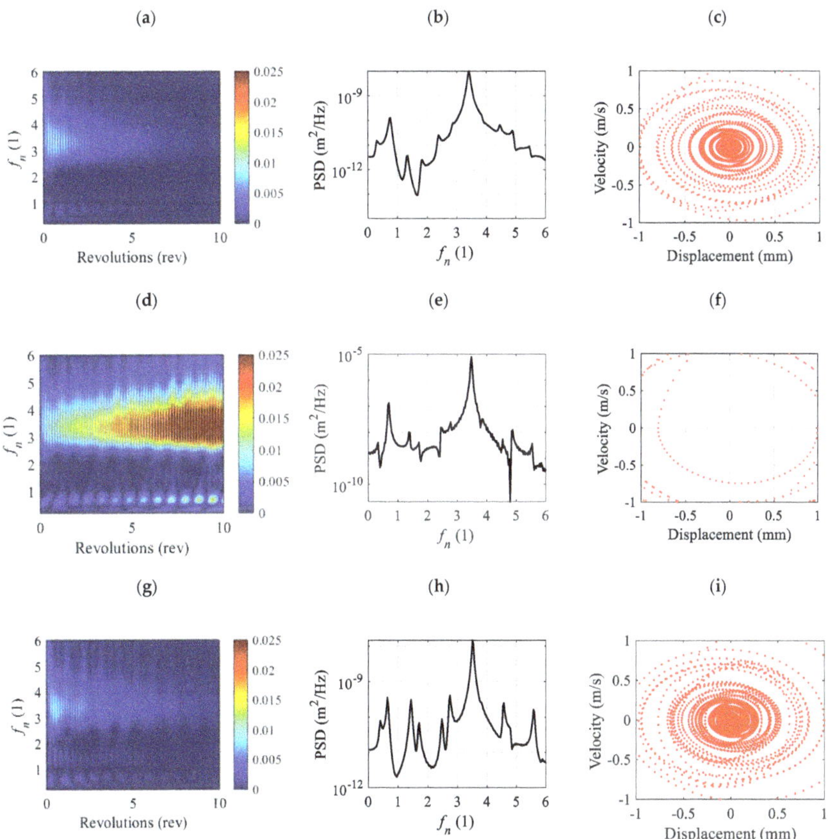

Figure 7. Analysis of cutting conditions A, B and C. Continuous wavelet transform (CWT) scalograms: (**a,d,g**); power spectral density (PSD): (**b,e,h**) and Poincaré maps (PM): (**c,f,i**) corresponds to the cutting conditions A, B and C respectively.

The PM diagram shown in Figure 7f exhibited a vibration far from zero. In fact, the PM diagram shows that the vibration amplitude grows exponentially because our equation of motion did not consider nonlinear effects such as those that appeared when the tool lost contact with the workpiece. Both cutting conditions A and B agreed with the stability boundaries in Figure 6. Now, the cutting conditions B were used but with a multivariable tool, which was referred to as cutting conditions C. The CWT plotted in Figure 7g described completely different results since there were no single dominant frequencies in comparison with cutting conditions A, but appeared several frequencies around $f_n = 3.2$ and close to $f_n = 1$ that reduced in intensity with time, suggesting a stable cutting. Figure 7i illustrates how the vibration amplitude approached to zero when using a multivariable tool in contrast to the PM obtained for the regular tool and exhibited in Figure 7f. This can be explained by observing that there were several excitation frequencies due to the irregular pitch and helix angles that break a single excitation frequency avoiding regenerative chatter phenomena.

Figure 8 illustrates the CWT, PSD and PM for simulated vibrations using the multivariable tool with different axial depths denoted as cutting conditions D, E, F and G for the axial depths of cut of 2.3, 3.0, 8.55 and 18 mm respectively. Notice that a stable case C was already validated when the axial depth was 1.7 mm in Figure 7g–i that corresponded to cutting conditions under the stability boundaries shown in Figure 6. For case D, a transient cutting condition was chosen very close to the critical axial depth of the cut. It is interesting to point out that transition cutting conditions in the CWT scalogram shown in Figure 8a not only shows frequencies with higher intensity in comparison with the stable case B, but also presents shifted frequencies that varied in intensity every single revolution. This shifting suggests a marginally stable cutting condition that was confirmed by the PM illustrated in Figure 8c, where circular trajectories were described close to the center point.

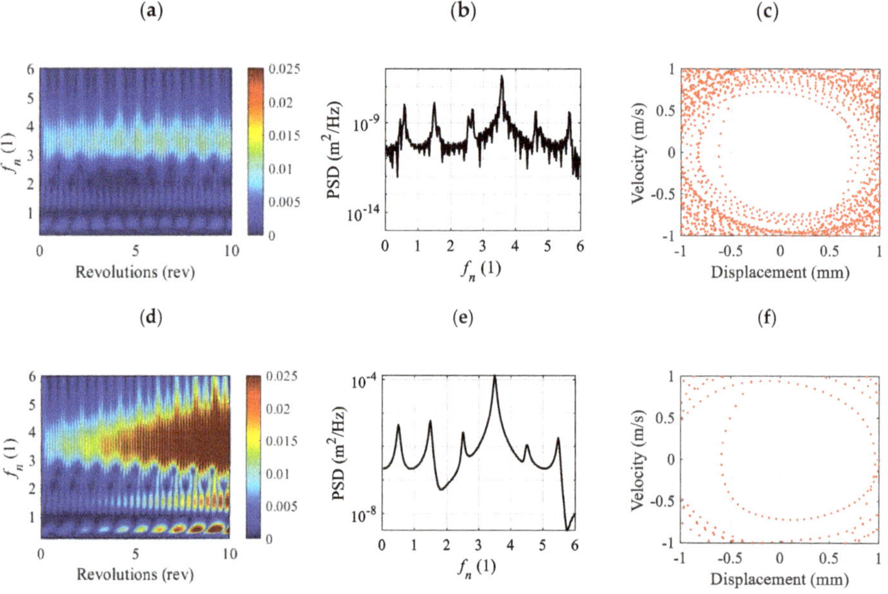

Figure 8. *Cont.*

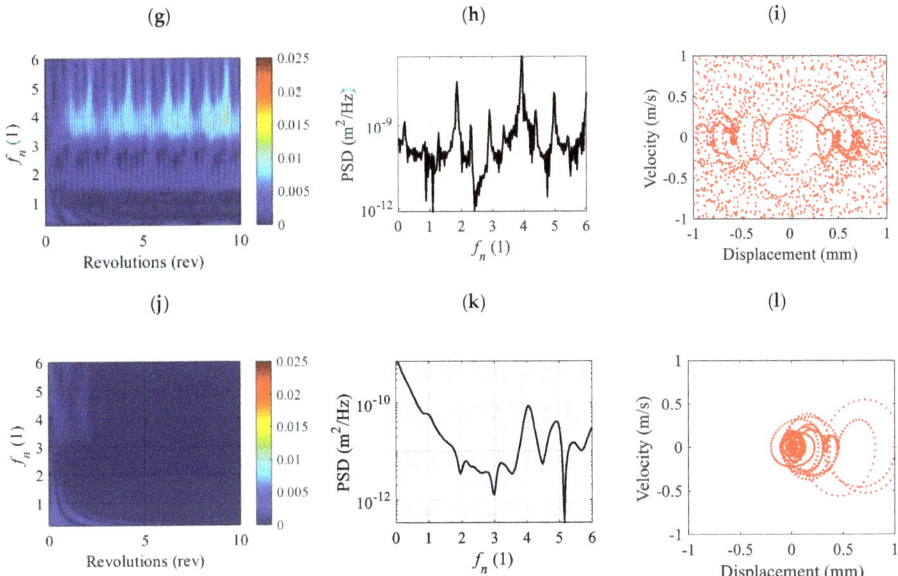

Figure 8. Analysis of cutting conditions D, E, F and G. CWT scalogram:s (**a,d,g,j**); PSD: (**b,e,h,k**) and PM: (**c,f,i,l**) corresponds to the cutting conditions D, E, F and G respectively.

Unstable vibrations that appeared for case E were because of the intensity of frequencies increased exponentially with time, see Figure 8d. Notice that other frequencies arose with time close to the values of $f_n = 0.5$ and $f_n = 1.5$. These frenquencies also occurred for cutting conditions D, which is an indication of the appearance of chatter phenomena. In contrast to Figure 7i for a stable case, Figure 8f exhibited few trajectories because the vibration amplitude was out of the range selected (±1 mm). The qualitative and quantitative dynamic behaviour due to cutting conditions F, and illustrated in Figure 8g-i, were classified as transition cutting behaviour. Here, a more severe shifting in frequencies was observed in the scalogram (Figure 8g). From Figure 8g, it is seen that drastic shifting occurred in the time domain in the range of normalized frequencies from 3.5 to 6. It was also evident in the PM showed in Figure 8i, that the amplitude of vibration remained below 1 mm during several revolutions of the tool but the amplitude of vibration never aproached to the center point, in contrast to the stable cutting condition C shown in Figure 7i in which the oscillation aplitudes aproached the center.

An interesting dynamic behaviour was observed in the milling cutting process when the cutting conditions were selected in the middle of the stable peninsula, above unstable cutting conditions such as E cutting conditions. The axial depth of the cut was increased from the unstable axial depth of cut of 3–18 mm, 6 times higher of the stable cutting condition C and 2 times higher than the unstable condition E. Since the vibration quickly decreased in a few revolutions no dominant frequencies appeared in the CWT and PSD failed to clearly identify a dominant frequency since the vibration amplitude decreased to zero after few revolutions, as confirmed by the PM shown in Figure 8l.

Figure 9 shows the normalized excitation frequencies that the multivariable tool produced for a fixed spindle speed of 2500 rpm. The total number of disks of 50 μm of thickness was grouped in sets of each millimeter in the axial direction. The waterfall plot in Figure 9 explains that a stable peninsula was formed above 11 mm because the workpiece was excited with several frequencies simultaneously. For instance, for a milling operation with the axial depth of cut of 1 mm (stable cutting), 80 discrete disks were cut with four normalized excitation frequencies values (3.3, 3.6, 4.5 and 5.1). On the other hand, when milling at 18 mm (stable cutting), there were 14 normalized excitation frequencies (3.30,

3.35, 3.39, 3.44, 3.49, 3.54, 3.60, 4.55, 4.64, 4.73, 4.82, 4.92, 5.02 and 5.13), most of them with at least 115 discrete disks.

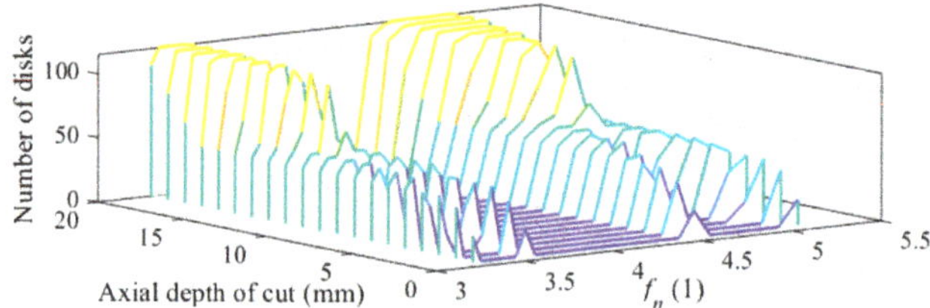

Figure 9. The number of discrete disks and discrete excitation frequencies as a function of the axial depth of cut for the multivariable tool.

5. Conclusions

In this work, quadratic and cubic polynomials were used to approximate the delayed terms of delay differential equations. Numerical simulations showed that using second- and third-order EMHPM improved the convergence rate and required less computational time when compared to the first-order EMHPM, and to semi-discretization and full-discretization methods, since fewer approximations or less discrete intervals were needed to reduce the computation time.

To further assess the applicability of the proposed method, the third-order EMHPM was used for determining the stability bounds in one-degree-of-freedom milling operation with a multivariable tool, demonstrating that the stability zone improved in comparison with a regular tool. For instance, at 2500 rpm the critical axial depth of the cut was 1.3 mm using the regular milling tool. However, using the multivariable tool, the critical axial depth of the cut was increased until 2.17 mm but more interesting, a stable zone appeared above 8.55 mm.

The CWT scalograms, PSD charts and PM were employed to validate the stability lobes found by using the third-order EMHPM for the multivariable tool. Numerical solutions confirmed the system dynamics behavior predicted by the third-order EMHPM.

Based on the above results, this paper provided evidence the third-order EMHPM could be used to study dynamic phenomena that appeared at higher axial depths of cut due to the multivariable design of the tool, which broke the excitation frequencies at a lower depth of cut.

Author Contributions: Conceptualization, J.d.l.L.S. and D.O.-T.; Methodology, J.d.l.L.S.; Resources, O.M.-R., A.E.-Z. and L.N.L.d.L.; Supervision, D.O.-T., A.E.-Z. and L.N.L.d.L.; Validation, J.d.l.L.S., D.O.-T. and G.U.; Writing—original draft, J.d.l.L.S. and D.O.-T.; Writing—review and editing, O.M.-R., G.U. and A.E.-Z. All authors have read and agreed to the published version of the manuscript.

Funding: This research was funded by Tecnológico de Monterrey through the Research Group of Nanotechnology for Devices Design, and by the Consejo Nacional de Ciencia y Tecnología de México (Conacyt), Project Numbers 242269, 255837, 296176, and National Lab in Additive Manufacturing, 3D Digitizing and Computed Tomography (MADiT) LN299129.

Acknowledgments: The authors acknowledge the technical assistance from Oscar Escalera at Tecnologico de Monterrey.

Conflicts of Interest: The authors declare no conflict of interest. The founding sponsors had no role in the design of the study; in the collection, analyses, or interpretation of data; in the writing of the manuscript, and in the decision to publish the results.

References

1. Mi, T.; Chen, N.; Stepan, G.; Takacs, D. Energy distribution of a vehicle shimmy system with the delayed tyre model. *IFAC-PapersOnLine* **2018**, *51*, 7–12. [CrossRef]
2. Orosz, G.; Stépán, G. Subcritical hopf bifurcations in a car-following model with reaction-time delay. *Proc. R. Soc. A Math. Phys. Eng. Sci.* **2006**, *462*, 2643–2670. [CrossRef]

3. Hu, H.; Zaihua, W. *Dynamics of Controlled Mechanical Systems with Delayed Feedback*; Springer Science & Business Media: Berlin, Germany, 2002; ISBN 9783642078392.

4. Altintas, Y. *Manufacturing Automation. Metal Cutting Mechanics, Machine Tool Vibrations, and CNC Design*; Cambridge University Press: New York, NY, USA, 2012.

5. Kuljanic, E.; Totis, G.; Sortino, M. Development of an intelligent multisensor chatter detection system in milling. *Mech. Syst. Signal Process.* **2009**, *23*, 1704–1718. [CrossRef]

6. Zhuo, Y.; Jin, H.; Han, Z. Chatter identification in flank milling of thin-walled blade based on fractal dimension. *Procedia Manuf.* **2020**, *49*, 150–154. [CrossRef]

7. Paul, S.; Morales-Menendez, R. Chatter mitigation in milling process using discrete time sliding mode control with type 2-fuzzy logic system. *Appl. Sci.* **2019**, *9*, 4380. [CrossRef]

8. Peng, C.; Wang, L.; Liao, T.W. A new method for the prediction of chatter stability lobes based on dynamic cutting force simulation model and support vector machine. *J. Sound Vib.* **2015**, *354*, 118–131. [CrossRef]

9. Li, D.; Cao, H.; Chen, X. Fuzzy control of milling chatter with piezoelectric actuators embedded to the tool holder. *Mech. Syst. Signal Process.* **2020**, *148*, 107190. [CrossRef]

10. Wan, S.; Li, X.; Su, W.; Yuan, J.; Hong, J.; Jin, X. Active damping of milling chatter vibration via a novel spindle system with an integrated electromagnetic actuator. *Precis. Eng.* **2019**, *57*, 203–210. [CrossRef]

11. Munoa, J.; Sanz-Calle, M.; Dombovari, Z.; Iglesias, A.; Pena-Barrio, J.; Stepan, G. Tuneable clamping table for chatter avoidance in thin-walled part milling. *CIRP Ann.* **2020**, *69*, 313–316. [CrossRef]

12. Wang, Y.; Wang, T.; Yu, Z.; Zhang, Y.; Wang, Y.; Liu, H. Chatter prediction for variable pitch and variable helix milling. *Shock Vib.* **2015**, *2015*. [CrossRef]

13. Mei, Y.; Mo, R.; Sun, H.; He, B.; Bu, K. Stability analysis of milling process with multiple delays. *Appl. Sci.* **2020**, *10*, 3646. [CrossRef]

14. Wan, M.; Zhang, W.H.; Dang, J.W.; Yang, Y. A unified stability prediction method for milling process with multiple delays. *Int. J. Mach. Tools Manuf.* **2010**, *50*, 29–41. [CrossRef]

15. Zhan, D.; Jiang, S.; Niu, J.; Sun, Y. Dynamics modeling and stability analysis of five-axis ball-end milling system with variable pitch tools. *Int. J. Mech. Sci.* **2020**, *182*, 105774. [CrossRef]

16. Slavicek, J. The effect of irregular tooth pitch on stability of milling. *Proc. 6th MTDR Conf.* **1965**, *1*, 15–22.

17. Budak, E. An analytical design method for milling cutters with nonconstant pitch to increase stability, Part I: Theory. *J. Manuf. Sci. Eng. Trans. ASME* **2003**, *125*, 29–34. [CrossRef]

18. Budak, E. An analytical design method for milling cutters with nonconstant pitch to increase stability, Part 2: Application. *J. Manuf. Sci. Eng. Trans. ASME* **2003**, *125*, 35–38. [CrossRef]

19. Altintas, Y.; Engin, S.; Budak, E. Analytical stability prediction and design of variable pitch cutters. *J. Manuf. Sci. Eng. Trans. ASME* **1999**, *121*, 173–178. [CrossRef]

20. Olgac, N.; Sipahi, R. Dynamics and stability of variable-pitch milling. *J. Vib. Control* **2007**, *13*, 1031–1043. [CrossRef]

21. Jin, G.; Zhang, Q.; Hao, S.; Xie, Q. Stability prediction of milling process with variable pitch cutter. *Math. Probl. Eng.* **2013**, *2013*. [CrossRef]

22. Comak, A.; Budak, E. Modeling dynamics and stability of variable pitch and helix milling tools for development of a design method to maximize chatter stability. *Precis. Eng.* **2017**, *47*, 459–468. [CrossRef]

23. Zatarain, M.; Muñoa, J.; Peigne, G.; Insperger, T. Analysis of the influence of mill helix angle on chatter stability. *CIRP Ann. Manuf. Technol.* **2006**, *55*, 365–368. [CrossRef]

24. Budak, E.; Altintas, Y. Analytical prediction of chatter stability in milling—Part I: General formulation. *Am. Soc. Mech. Eng. Dyn. Syst. Meas. Control* **1998**, *120*, 22–30. [CrossRef]

25. Insperger, T.; Muñoa, J.; Zatarain, M.; Peigné, G. Unstable islands in the stability chart of milling processes due to the helix angle. In Proceedings of the CIRP 2nd International Conference High Performance Cutting, Vancouver, BC, Canada, 12–13 June 2006.

26. Patel, B.R.; Mann, B.P.; Young, K.A. Uncharted islands of chatter instability in milling. *Int. J. Mach. Tools Manuf.* **2008**, *48*, 124–134. [CrossRef]

27. Sims, N.D.; Mann, B.; Huyanan, S. Analytical prediction of chatter stability for variable pitch and variable helix milling tools. *J. Sound Vib.* **2008**, *317*, 664–686. [CrossRef]

28. Turner, S.; Merdol, D.; Altintas, Y.; Ridgway, K. Modelling of the stability of variable helix end mills. *Int. J. Mach. Tools Manuf.* **2007**, *47*, 1410–1416. [CrossRef]

29. Yusoff, A.R.; Sims, N.D. Optimisation of variable helix tool geometry for regenerative chatter mitigation. *Int. J. Mach. Tools Manuf.* **2011**, *51*, 133–141. [CrossRef]
30. Dombovari, Z.; Stepan, G. The effect of helix angle variation on milling stability. *J. Manuf. Sci. Eng. Trans. ASME* **2012**, *134*. [CrossRef]
31. Huang, J.; Deng, P.; Li, H.; Wen, B. Stability analysis for milling system with variable pitch cutters under variable speed. *J. Vibroeng.* **2019**, *21*, 331–347. [CrossRef]
32. Cai, S.; Yao, B.; Feng, W.; Cai, Z.; Chen, B.; He, Z. Milling process simulation for the variable pitch cutter based on an integrated process-machine model. *Int. J. Adv. Manuf. Technol.* **2020**, *106*, 2779–2791. [CrossRef]
33. Olvera, D.; Elías-Zúñiga, A.; López De Lacalle, L.N.; Rodríguez, C.A. Approximate solutions of delay differential equations with constant and variable coefficients by the enhanced multistage homotopy perturbation method. *Abstr. Appl. Anal.* **2015**, 1–12. [CrossRef]
34. Compeán, F.I.; Olvera, D.; Campa, F.J.; López De Lacalle, L.N.; Elías-Zúñiga, A.; Rodríguez, C.A. Characterization and stability analysis of a multivariable milling tool by the enhanced multistage homotopy perturbation method. *Int. J. Mach. Tools Manuf.* **2012**, *57*, 27–33. [CrossRef]
35. Sims, N.D. Fast chatter stability prediction for variable helix milling tools. *Proc. Inst. Mech. Eng. Part C J. Mech. Eng. Sci.* **2016**, *230*, 133–144. [CrossRef]
36. Otto, A.; Rauh, S.; Ihlenfeldt, S.; Radons, G. Stability of milling with non-uniform pitch and variable helix Tools. *Int. J. Adv. Manuf. Technol.* **2017**, *89*, 2613–2625. [CrossRef]
37. Niu, J.; Ding, Y.; Zhu, L.M.; Ding, H. Mechanics and multi-regenerative stability of variable pitch and variable helix milling tools considering runout. *Int. J. Mach. Tools Manuf.* **2017**, *123*, 129–145. [CrossRef]
38. Olvera, D.; Urbikain, G.; Elías-Zuñiga, A.; López de Lacalle, L.N. Improving stability prediction in peripheral milling of Al7075T6. *Appl. Sci.* **2018**, *8*, 1316. [CrossRef]
39. Iglesias, A.; Dombovari, Z.; Gonzalez, G.; Munoa, J.; Stepan, G. Optimum selection of variable pitch for chatter suppression in face milling operations. *Materials* **2018**, *12*, 112. [CrossRef] [PubMed]
40. Guo, Y.; Lin, B.; Wang, W. Optimization of variable helix cutter for improving chatter stability. *Int. J. Adv. Manuf. Technol.* **2019**, *104*, 2553–2565. [CrossRef]
41. Hashim, I.; Chowdhury, M.S.H. Adaptation of homotopy-perturbation method for numeric-analytic solution of system of ODEs. *Phys. Lett. Sect. A* **2008**, *372*, 470–481. [CrossRef]
42. Puma-Araujo, S.D.; Olvera-Trejo, D.; Martínez-Romero, O.; Urbikain, G.; Elías-Zúñiga, A.; Lacalle, L.N.L. de Semi-active magnetorheological damper device for chatter mitigation during milling of thin-floor components. *Appl. Sci.* **2020**, *10*, 5313. [CrossRef]
43. Insperger, T.; Stépán, G. Updated semi-discretization method for periodic delay-differential equations with discrete delay. *Int. J. Numer. Methods Eng.* **2004**, *61*, 117–141. [CrossRef]
44. Insperger, T. Full-discretization and semi-discretization for milling stability prediction: Some comments. *Int. J. Mach. Tools Manuf.* **2010**, *50*, 658–662. [CrossRef]
45. Ding, Y.; Zhu, L.M.; Zhang, X.J.; Ding, H. A full-discretization method for prediction of milling stability. *Int. J. Mach. Tools Manuf.* **2010**, *50*, 502–509. [CrossRef]
46. Lee, E.T.; Eun, H.C. Structural damage detection by power spectral density estimation using output-only measurement. *Shock Vib.* **2016**, *2016*. [CrossRef]
47. Fries, R.H. *Fundamentals of Vibrations*; Waveland Press: Long Grove, IL, USA, 2000; ISBN 9781482270372.

Publisher's Note: MDPI stays neutral with regard to jurisdictional claims in published maps and institutional affiliations.

Article

Natural Frequency Prediction Method for 6R Machining Industrial Robot

Jiabin Sun [1,*], Weimin Zhang [1] and Xinfeng Dong [2]

[1]   School of Mechanical Engineer, Tongji University, ShangHai 201804, China; iamt@tongji.edu.cn

[2]   College of Energy and Mechanical Engineering, Shanghai University of Electric Power, ShangHai 201303, China; laile_sd@shiep.edu.cn

*   Correspondence: 13tjsunjb@tongji.edu.cn

Received: 10 October 2020; Accepted: 13 November 2020; Published: 17 November 2020

Abstract: The industrial robot machining performance is highly dependent on dynamic behavior of the robot, especially the natural frequency. This paper aims at introducing a method to predict the natural frequency of a 6R industrial robot at random configuration, for improving dynamic performance during robot machining. A prediction model of natural frequency which expresses the mathematical relation between natural frequency and configuration is constructed for a 6R robot. Joint angles are used as input variables to represent the configurations in the model. The quantity and range of variables are limited for efficiency and practicability. Then sample configurations are selected by central composite design method due to its capacity of disposing nonlinear effects, and natural frequency data is acquired through experimental modal test. The model, which is in form of regression equation, is fitted and optimized with sample data through partial least square (PLS) method. The proposed model is verified with random configurations and compared with the original model and a model fitted by least square method. Prediction results indicate that the model fitted and optimized by PLS method has the best prediction ability. The universality of the proposed method is validated through implementation onto a similar 6R robot.

Keywords: machining robot; natural frequency prediction; model optimization; dynamic performance

---

1. Introduction

Due to the rapid development of industrial automation, the global amount of industrial robot keeps increasing significantly in recent years [1], and new applications attract more attention [2–5]. Because of the multiple degree of freedom, large workspace, and relative low cost, industrial robot becomes a potential alternative of traditional machine tool in hard material drilling [6], boring [7], and milling [8,9] processes, which are not maturely and widely applied.

For machine tool, in the whole workspace, the stiffness and dynamic behavior nearly remain steady, so machining stability and chatter suppression, which are concerned with multiple factors, are research emphasis [10–12]. However, for industrial robot, current researches focus on the characteristic of robot itself. Common industrial robot is of open-loop articulated serial structure with six revolute joints (6R); therefore, the stiffness and dynamic performance are much inferior to machine tool, which results in vibration issues and poor surface quality in machining process [13]. Static stiffness issues are studied to avoid the robot deformation in machining, and several stiffness indexes are proposed to evaluate the static stiffness of robot in the workspace [14–17]. On the other hand, dynamic behavior of robot is studied to evaluate the capacity of dealing with time-variant load in machining process, which may help to suppress vibration [18,19]. As 6R robot realizes movement through adjusting joint angles, dynamic behavior keeps varying with configurations during motion process. For that reason, research object is gradually expanded from one configuration to the whole workspace. Bisu et al. [20] selected three configurations from a linear movement path of a 6R robot with interval of 50 cm for dynamic behavior study and vibration analysis. It was found that different natural frequencies led to dynamic performance variation. Mousavi et al. [21] performed a similar four-point-study and analyzed the change of stability region through modal information. Several configurations may present a linear movement, but the usage of robot should not be such limited. Karim et al. [22] executed modal test with 63 measured points for 23 configurations, through which mode shapes were obtained and approximate distribution of first two step natural frequencies in a plane was depicted. Natural frequency is the vital dynamic parameter in robot machining, but massive tests are inefficient although the accuracy is well ensured. A method was proposed by Glogowski et al. [23] to predict natural frequency in the whole workspace efficiently, in which sample configurations were selected by design of experiment (DoE) method and prediction model was fitted by least square (LS) method with the natural frequency data of sample configurations, but the prediction deviation reached up to 8.9%, making the method inadequate for practical application.

The low order natural frequencies of 6R robot are relatively small and the first order is merely about 10 Hz, which is covered in frequency range of most machining processes, so 6R robot tends to forced vibration during machining processes [24]. To obtain basis for vibration suppression, structure optimization and path planning, low order natural frequencies are particularly concerned among dynamic parameters. It is proposed that low order natural frequencies of robot are decided by the configurations, while the high order natural frequencies are mainly related to the machining system, which do not vary with robot configurations [25], this phenomenon is demonstrated in Figure 1.

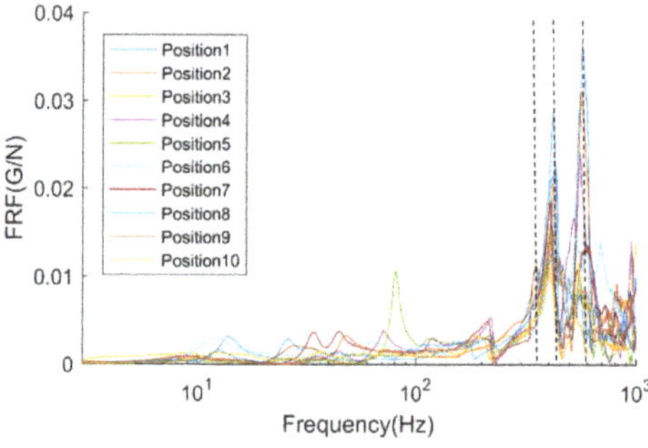

Figure 1. Frequency Response Functions of different robot configurations.

Thus, a method to predict natural frequency of 6R industrial robot efficiently and precisely is proposed in this paper. As shown in Figure 2, sample configurations are selected to present the practical workspace (variable range) for fast prediction, and a regression equation fitted and optimized by partial least square (PLS) method is used as prediction model for higher accuracy. A 6R robot, ABB IRB4600, is used to illustrate and validate the method. Another ABB IRB6400 is used verify the universality of proposed method.

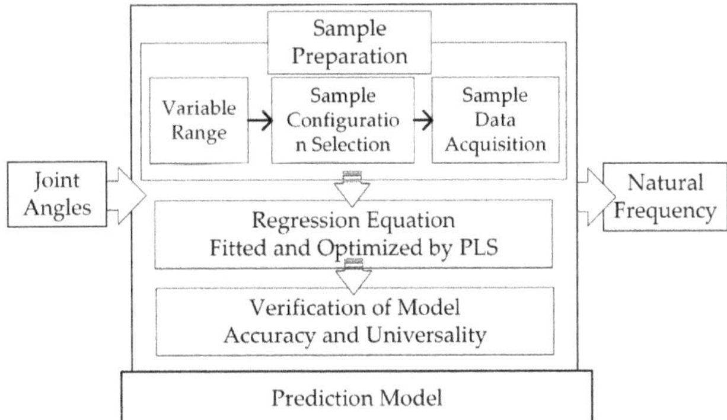

Figure 2. Prediction method of robot natural frequency.

2. Materials and Methods

2.1. Sample Preparation

The axes of 6R robot are noted as a_1, a_2, ... , a_6, corresponding joint angles are q_1, q_2, ... , q_6. The mapping relation between natural frequency (f_F) and configuration can be expressed as

$$f_F = f(q_1, q_2, \ldots, q_6), \tag{1}$$

With samples, a regression equation can be fitted to represent the mapping relation. However, it is unnecessary and inefficient to include the entire workspace of the robot as the configurations near the boundary could not be adopted in machining process. Therefore, range of variables is limited before selecting samples through DoE method. Then natural frequency information of each sample configuration is acquired through modal test.

2.1.1. Variable Range Determination

a_1, which is the basic rotation axis of the robot structure, and a_6, which is a 360° rotation axis controlling the flange, are proved to be nonsignificant in changing natural frequency by preliminary researches on 6R robots including KUKA KR60, ABB IRB6400, and IRB 4600. As low order natural frequencies is mainly decided by robot configuration, none end-effector is fixed on the flange in above cases for general research. Hereby, q1 and q6 are excluded from model ($q_1 = q_6 = 0°$ in practice) for the simplification, and q_2–q_5 are applied as variables. Prediction model can be adapted to

$$f_F = f(q_2, q_3, q_4, q_5), \tag{2}$$

To ensure the efficiency of model, partial workspace that may cause movement interference in machining process is rejected by simulation software. The variable range is finally determined and shown in Figure 3, where the solid area indicates the new efficient workspace, and the dashed area represents the original workspace.

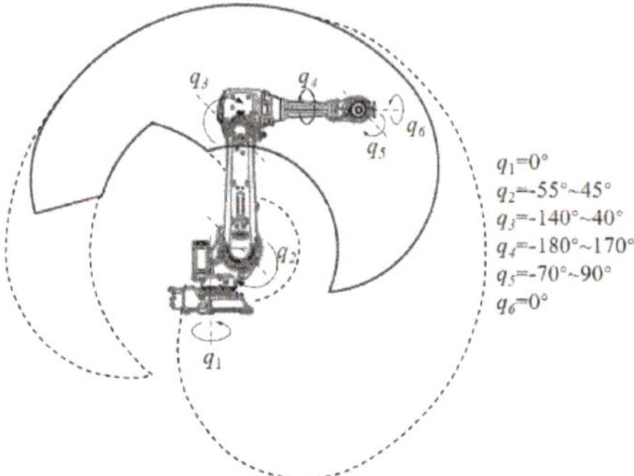

Figure 3. Partial variable range.

2.1.2. Sample Configuration Selection

DoE method has advantages in study influence of different factors, especially for basic research of robot machining, e.g., Simoes et al. [26] used DoE to study influence of robot milling parameters. Different types of selection methods are compared in pilot study including three-level full factorial design (3kD), orthogonal experimental design, Box-Behnken design, and central composite design (CCD). The prediction performances of 3kD and CCD are nearly matched and superior to the other two, while CCD demands a much smaller sample amount [23]. Meanwhile CCD owns conspicuous ability in disposing the nonlinear influence that joint angles may lead to natural frequency. Therefore, CCD method is adopted to select sample configurations in this paper, corresponding factors and levels are shown in Table 1.

Table 1. Factors and Levels of central composite design (CCD).

Factors Levels	q_2 (Deg)	q_3 (Deg)	q_4 (Deg)	q_5 (Deg)
−1	−55	−140	−180	−70
−0.5	−30	−95	−92.5	−30
0	−5	−50	−5	10
+0.5	20	−5	82.5	50
1	45	40	170	90

2.1.3. Sample Data Acquisition

Experimental modal test (EMT) is applied so that the natural frequency data of sample configurations can be identified accurately. According to principle of modal test, any element of frequency response function (FRF) can be written as

$$h_{ij}(\omega) = \sum_{k=1}^{n} \left(\frac{a_{ijk}}{(j\omega - p_k)} + \frac{a_{ijk}^*}{(j\omega - p_k^*)} \right), \tag{3}$$

which presents the response of i point caused by the excitation at j point. a_{ijk} and a_{ijk}^* are related to mode shapes, their values are depending on relevant combination of excitation and response. p_k and p_k^* contain the information of natural frequency and damping, which are global characteristics independent of excitation and response combinations. Theoretically, one combination of excitation and response is enough to identify all the natural frequencies, provided the excitation and response points avoid the nodes.

The sample configurations selected by CCD design vary greatly, for the convenience of implementing all the trials in the CCD design, hammer excitation is chosen, and a three-direction accelerometer is fixed on the flange of robot to acquire response signal. In this way, the extra mass brought by EMT device can be negligible as the accelerometer is small and light compared to the robot. Excitations are generated from three directions for each trial, which are corresponding to the three directions of the accelerometer, as shown in Figure 4. Then three FRFs will be obtained for natural frequency identification, avoiding incomplete excitation or excitation on the nodes. Moreover, for each direction, five groups of excitation and response signals are acquired to decrease the signal-to-noise ratio through average algorithm. With FRFs the natural frequencies are preliminarily identified, then complex modal indicating function is used to confirm the accurate values of natural frequencies.

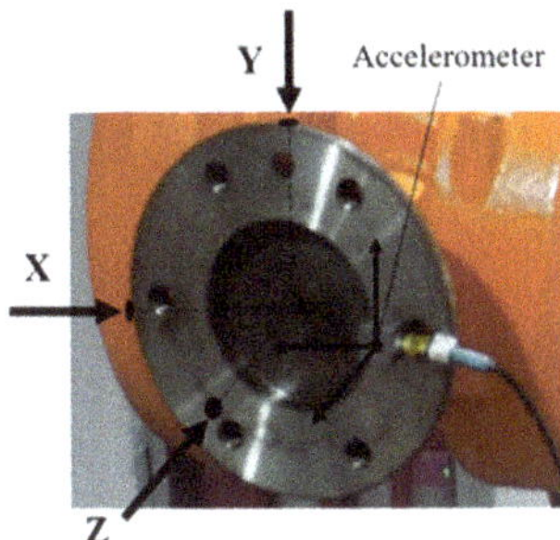

Figure 4. Directions of accelerometer and excitations.

With the ABB IRB4600 industrial robot being the object, a EMT system for sample data acquisition is set up, including a LIXIE hammer, a PCB three-direction accelerometer, an ECON device for vibration signal processing and a PC (as shown in Figure 5). The configuration of robot is adjusted according to the CCD design in turn. All the identified natural frequency information will be used as the sample data to construct a prediction model.

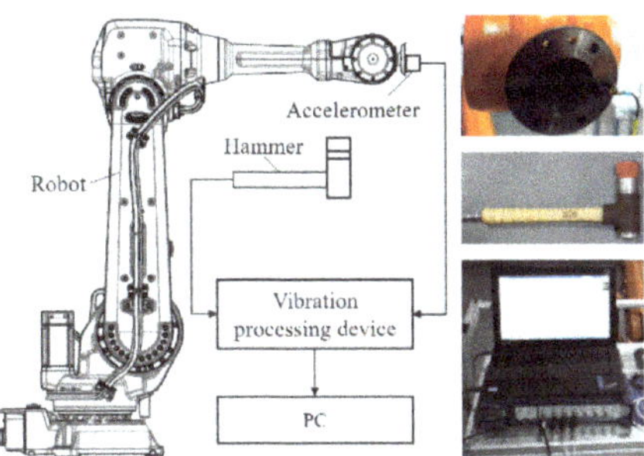

Figure 5. The setup of experimental modal test (EMT) system.

2.2. Prediction Model Construction and Optimization

Because of the serial structure, the relation between configuration and joint angles could not be linear. Thus, Equation (2) is fitted to a general second order equation in this case according to normal engineering application. PLS method is adopted to solve the issues of nonlinear influence and multi-correlation between joint angles in regression equation fitting. Meanwhile, the fitting error caused by the limitation of sample quantity can be remarkably reduced, due to the advantage of PLS method in coping with small sample quantity. In addition, PLS has good interpretability for the terms of regression equation, which can be used as optimization criterion.

2.2.1. PLS Method Fitting Process

The principle of PLS method in multivariate regression fitting is to extract principal components t and u respectively from independent variables X and dependent variables Y while retaining the most correlation, which can be mathematically described as

$$\text{Cov}(t, u) = \sqrt{\text{Var}(t)\text{Var}(u)}\, r(t, u) \to \max, \tag{4}$$

The process of regression equation fitting by PLS method [27] is briefly introduced as follows.

1. Data standardization.
2. First principal component calculation.
3. Principal component validation.

Current principal component t_h is validated by following indexes:

$$R_Y^2(\text{cum}) = \sum_{h=1}^{m} r^2(Y, t_h), \tag{5}$$

$$Q_h^2 = 1 - \frac{S_{SPE,h}}{S_{SE,h}}, \tag{6}$$

$$Q_h^2(\text{cum}) = 1 - \prod_{k=1}^{h} \frac{S_{SPE,k}}{S_{SE,k}}, \tag{7}$$

where $S_{SPE,h}$ is the square sum of prediction error of Y_{h-1}, $S_{SE,h}$ is square sum of deviation of Y_{h-1}.
$R_Y^2(\text{cum})$ indicates the ability of principal component to explain the variability of Y_{h-1}, and its upper limit value is 1. Q_h^2 represents the contribution of t_h to the model. When it is more than $(1{-}0.95)^2 = 0.0975$, t_h has significant effect on prediction model. Meanwhile, $Q_h^2(\text{cum})$ increases, meaning that the comprehensive significance in explaining Y is enhanced.

4. Subsequent principal components calculation. The residual matrixes from Step c. are continued to compute new principal components, until t_{Q+1} deduces $Q_h^2(\text{cum})$. The principal components before t_{Q+1} are validated ones.
5. Regression equation

$$Y_0 = t_1 r_1^T + t_2 r_2^T + \ldots + t_Q r_Q^T + Y_Q, \tag{8}$$

6. Data reduction. The final regression equation for prediction can be obtained through the reverse process of data standardization.

2.2.2. Interpretability of PLS Method for Regression Equation

Variable importance in projection (VIP), which indicates the importance of each item in regression equation, is defined as

$$\text{VIP}_j = \sqrt{\frac{p}{R_Y^2(\text{cum})} \sum_{h=1}^{m} r^2(Y, t_h) w_{hj}^2}, \tag{9}$$

where w_{jh} is the jth component of w_h.
The larger VIP_j is, the more important x_j is in constructing Y. The effects of different items can be compared quantitatively through VIP values, so that some redundant items can be rejected to realize regression equation optimization.

2.2.3. Prediction Model Construction

The sample data from Section 2.1.3 is utilized to construct prediction model, taking the first order natural frequency of ABB IRB4600 robot as example.

As mentioned, the general regression equation concludes q_2, q_3, q_4, q_5, and their cross items and square items. The general equation is marked as M. After the second principal component is extracted, validation indicates are shown in Table 2. The second principal component deduces $Q_h^2(\text{cum})$, meaning that comprehensive ability is weakened. Though $R_Y^2(\text{cum})$ increases, the second principal component is still abandoned.

Table 2. Principal component validation of M.

h	$R_Y^2(\text{cum})$	Q_h^2	$Q_h^2(\text{cum})$
1	0.888	0.639	0.639
2	0.962	−0.171	0.603

2.2.4. Prediction Model Optimization

$Q_h^2(\text{cum})$ of M is merely 0.639, which means M is inadequate to represent f_F. To optimize the model, VIP values of items in M is calculated and illustrated in Figure 6.

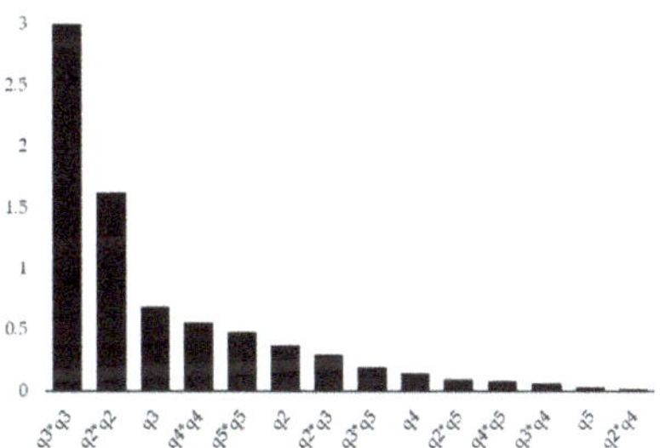

Figure 6. VIP values of items in M.

All the one-degree items should be retained as basic items. All the square items are important enough to stay in the model because of their high VIP values. As VIP values of cross items are relatively low, different combination groups should be tested. A new model M_1 is constructed by cutting off cross items except q_2q_3, because q_2q_3 has the largest VIP value among cross items; meanwhile, q_2 and q_3 are most two important variables according to VIP values. Based on M_1, the other five cross items are added respectively in descending order of their VIP values. The new five models are marked as M_2–M_6. Validation results of above models are shown in Table 3. All the six models are constructed with two principal components. $R_Y^2(\text{cum})$ and $Q_h^2(\text{cum})$ values of the six models are all improved obviously compared to M. M_2, formed by adding q_3q_5 to M_1, has the best $R_Y^2(\text{cum})$ and $Q_h^2(\text{cum})$ values. M_3–M_6, which get lower $R_Y^2(\text{cum})$ and $Q_h^2(\text{cum})$ values compared to M_1, are abandoned.

Based on M_2, four new models are introduced by adding q_2q_4, q_3q_4, q_2q_5, and q_4q_5 in turn. However, the largest $Q_h^2(\text{cum})$ only reaches up to 0.764. All the other combination groups of above four cross items are add to M_2 for test, but none of them can acquire better validation result. Thus, the optimized model is finally confirmed to be M_2. The coefficients of items are listed in Table 4 together with M.

Table 3. Validation results of M_1–M_6.

Model	h	$R_Y^2(\text{cum})$	$Q_h^2(\text{cum})$
M_1	2	0.958	0.776
M_2	2	0.960	0.778
M_3	2	0.958	0.764
M_4	2	0.958	0.762
M_5	2	0.958	0.723
M_6	2	0.958	0.717

Table 4. Coefficients of items in M and M_2.

Items	Coefficient/M	Coefficient/M_2
Constant	1.138×10^1	1.086×10^1
q_2	8.746×10^{-3}	1.214×10^{-2}
q_3	5.987×10^{-2}	6.611×10^{-2}
q_4	-7.532×10^{-5}	7.873×10^{-4}
q_5	1.374×10^{-4}	-3.785×10^{-3}
q_2^2	8.821×10^{-4}	1.179×10^{-3}
q_3^2	5.322×10^{-4}	5.894×10^{-4}
q_4^2	-2.464×10^{-5}	1.242×10^{-5}
q_5^2	-1.018×10^{-4}	7.342×10^{-5}
$q_2 q_3$	1.171×10^{-4}	1.253×10^{-4}
$q_2 q_4$	4.323×10^{-6}	-
$q_2 q_5$	4.224×10^{-5}	-
$q_3 q_4$	-7.206×10^{-6}	-
$q_3 q_5$	-4.939×10^{-5}	-5.285×10^{-5}
$q_4 q_5$	1.099×10^{-5}	-

3. Results

3.1. Prediction Model Verification and Analysis

3.1.1. Model Verification with Random Configurations

Sixteen groups of q_2–q_5 values are picked randomly from the variable range defined in Section 2.1.1, through which 16 corresponding robot configurations are defined for model verification. The simulation images of the 16 configurations are reranked by joint angles q_2 and q_3 as shown in Figure 7, and the measured natural frequency is listed in the bottom left corner of each image. The configurations are nearly uniformly distributed in the variable range, which ensures the comprehensiveness of verification objects and rationality of verification result.

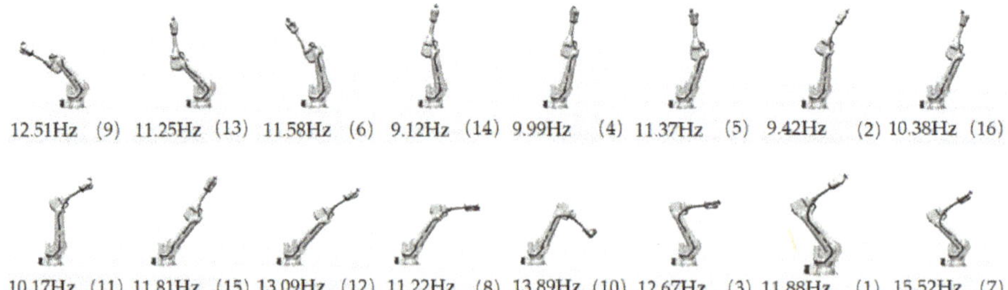

Figure 7. Random configurations for model verification.

The first order natural frequencies of the 16 configurations are obtained by EMT. Then, original model M and optimized model M_2 are utilized to predict the first order natural frequencies. The deviation between prediction value and measured value is defined as prediction error, the details are listed in Table 5; Table 6.

Table 5. Prediction error of M.

No.	1	2	3	4	5	6	7	8
Error	0.20	0.36	0.22	0.11	0.22	0.45	1.59	0.26
Relative error %	1.71	3.79	1.76	1.10	1.90	3.88	10.24	2.34
No.	9	10	11	12	13	14	15	16
Error	0.54	1.43	0.87	1.04	1.46	1.78	1.47	0.12
Relative error %	4.34	10.30	8.40	8.27	12.56	19.54	12.39	1.18
Average error	0.76			Average relative error %			6.48	

Table 6. Prediction error of M_2.

No.	1	2	3	4	5	6	7	8
Error	0.30	0.10	0.19	0.46	0.44	0.44	0.30	0.32
Relative error %	2.51	1.04	1.47	4.55	3.85	3.81	1.92	2.88
No.	9	10	11	12	13	14	15	16
Error	0.51	0.44	0.20	0.51	0.42	0.24	0.11	0.15
Relative error %	4.12	3.14	1.90	4.01	3.56	2.68	0.88	1.48
Average error	0.32			Average relative error %			2.74	

The average error and average relative error of optimized model M_2 are both much lower than those of original model M. The average relative error of M_2 is less than 5% which brings more reliability. On the other hand, seven high relative errors (more than 5%) which are underlined show up in the prediction process of M, and the maximum error reaches up to 1.78 Hz. While M_2 has no predictions with over 5% relative error, and the maximum error is 0.51 Hz which is acceptable. Obviously, optimized model M_2 has better prediction ability.

A model (M_{LS}) fitted by least square (LS) method with the same sample data is taken as the contrast. Moreover, the prediction errors are displayed in Table 7. Average error and average relative error of M_{LS} are both higher than those of M and M_2, that is, PLS method have better performance in construct prediction model of natural frequency than LS method. In conclusion, model fitted and optimized by PLS method has the best prediction ability.

Table 7. Prediction error of M_{LS}.

No.	1	2	3	4	5	6	7	8
Error	0.51	0.45	0.51	0.79	1.16	1.37	0.52	0.86
Relative error %	4.32	4.77	4.01	7.91	10.30	11.81	4.66	7.68
No.	9	10	11	12	13	14	15	16
Error	1.49	0.64	0.55	0.40	1.94	0.36	0.46	0.43
Relative error %	11.92	6.32	4.86	4.38	16.44	3.29	4.43	5.89
Average error	0.78			Average relative error %			7.07	

3.1.2. Model Construction Quality Analysis

In PLS method, when model is constructed by principal components, information in the last residual matrix is ignored, causing the error between fitted model and original data. Standardized distance between sample point and model can be used to evaluate the construction quality. $(DModX,N)_i$

or $(DModY,N)_i$ is the ratio of construct quality of the *i*th sample point and the average construct quality. When the ratio is less than two, the construction for the *i*th sample point is reasonable. Construction quality of M_2 is shown in Figure 8. $(DModX,N)_i$ and $(DModY,N)_i$ are all less than two, that means the construction of M_2 is rational.

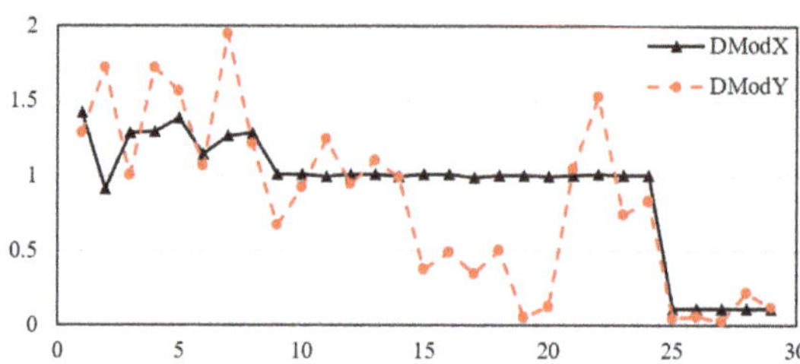

Figure 8. Construction quality of M2.

3.2. Method Universality Verification

To testify the universality of the natural frequency prediction method proposed in this paper, it is completely applied to an ABB IRB6400 industrial robot (as shown in Figure 9). Sample data are obtained according to Chapter 2.1 as the structure is similar to IRB4600. PLS is used to fit and optimize the prediction model M_{6400}. In this case, q_2q_4, q_2q_5, and q_4q_5 are abandoned according to VIP values. Validation results can be seen in Table 8, values of $R_Y^2(\text{cum})$ and $Q_h^2(\text{cum})$ indicate that the fitting is rational. Construction quality is eligible as shown in Figure 10. The result of model verification through random configurations is shown in Table 9, average relative error is merely 2.982%, indicating the good prediction performance.

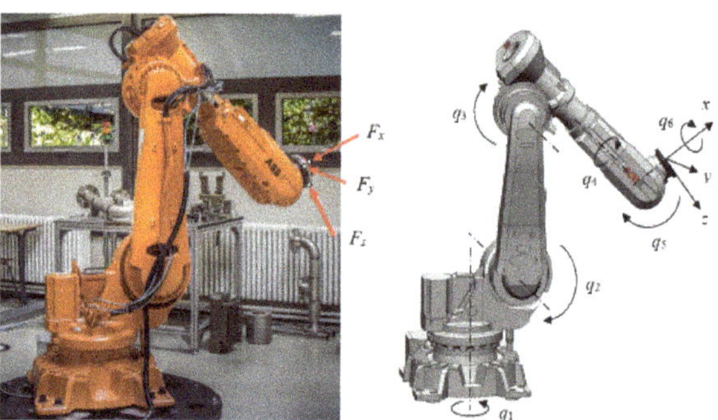

Figure 9. ABB IRB6400 industrial robot.

Table 8. Validation results of M_{6400}.

h	$R_Y^2(\text{cum})$	Q_h^2	$Q_h^2(\text{cum})$
1	0.801	0.720	0.720
2	0.952	0.546	0.874

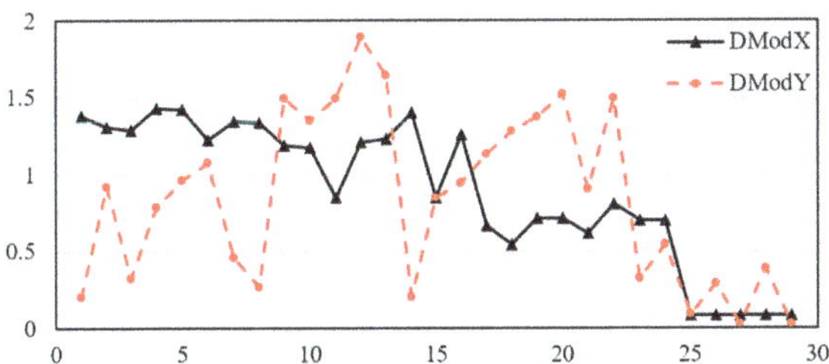

Figure 10. Construction quality of M_{6400}.

Table 9. Prediction error of M_{6400}.

No.	1	2	3	4	5	6	7	8
Error	0.35	0.43	0.13	0.47	0.23	0.23	0.07	0.19
Relative error %	2.94	3.38	1.41	4.16	1.97	1.97	0.44	1.66
No.	9	10	11	12	13	14	15	16
Error	0.15	0.23	0.25	0.61	0.78	0.58	0.35	0.34
Relative error %	1.56	2.05	2.74	4.81	7.19	5.59	3.11	2.78
Average error		0.33	Average relative error %				2.98	

The two models constructed for two different robots through the method proposed in this paper both own good prediction ability. That is, the universality of the proposed method is testified.

4. Discussion

Once the prediction model of natural frequency is constructed and verified, the prediction result can be used as optimization parameter to improve the machining performance. An application case of the prediction model is explained as follows.

a_2 and a_3 control the two longest links, by which the basic configuration of robot is determined. Through VIP values, items involving q_2 and q_3 are demonstrated to act dominated role in prediction model. a_4 and a_5 decide the direction of robot flange, which affect the configuration less than a_2 and a_3. Taking M_2 for an example, the influence of q_2 and q_3 on natural frequency is specifically studied and illustrated in Figure 11. When $q_2 = -55°$ and $q_3 = 40°$ (P_{max}), the maximum first order natural frequency is obtained, and the minimum appears when $q_2 = -2°$ and $q_3 = -56°$ (P_{min}).

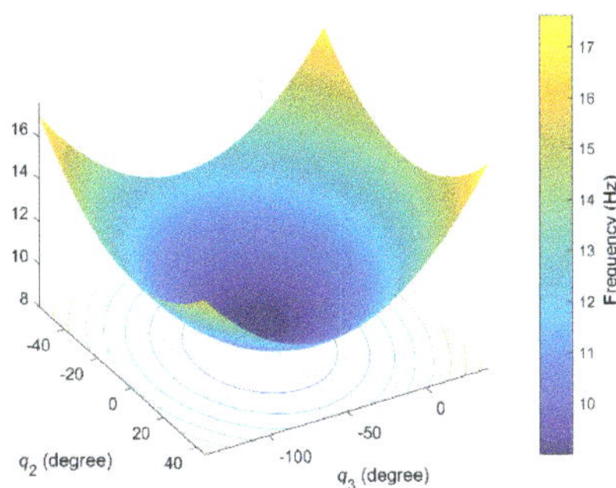

Figure 11. Variation of natural frequency with q_2 and q_3.

Three configurations near P_{max}, P1 ($q_2 = -48°$, $q_3 = 50°$), P2 ($q_2 = -38°$, $q_3 = 45°$), and P3 ($q_2 = -28°$, $q_3 = 40°$) are chosen for milling test (as shown in Figure 12. The first order natural frequencies decrease in turn from P1 to P3 according to model M_2. The setup of milling test is shown in Figure 13. High speed milling with short and straight path are executed, so that robot configuration can be ignored in one milling path. Milling parameters are as follows: the spindle rotation frequency is 800 Hz, f is 2.4 mm/s, a_e is 1 mm, and a_p is 4 mm. Acceleration signals and milled surface are analyzed.

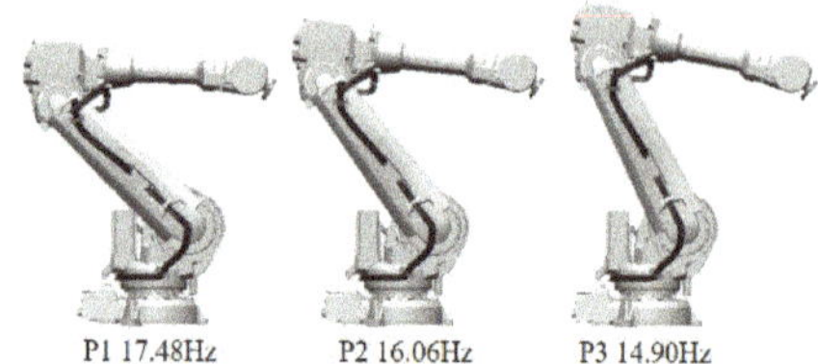

Figure 12. Chosen robot configurations for milling test.

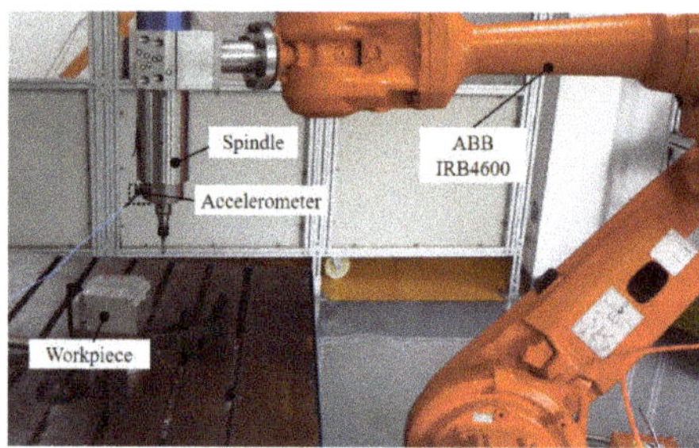

Figure 13. Milling test system setup.

Milling acceleration signals are treated with short-time fourier transform (STFT), spectrograms of the range 5–100 Hz are displayed in Figure 14. For all three configurations, several peaks are conspicuous around 50–90 Hz during the whole milling process. In low frequency stage, a peak about 16 Hz can be found in P2 and a 15 Hz peak for P3, which are corresponding to their predicted natural frequencies. That no conspicuous can be found in P1, may because the certain frequency is not significantly impacted by the milling parameters.

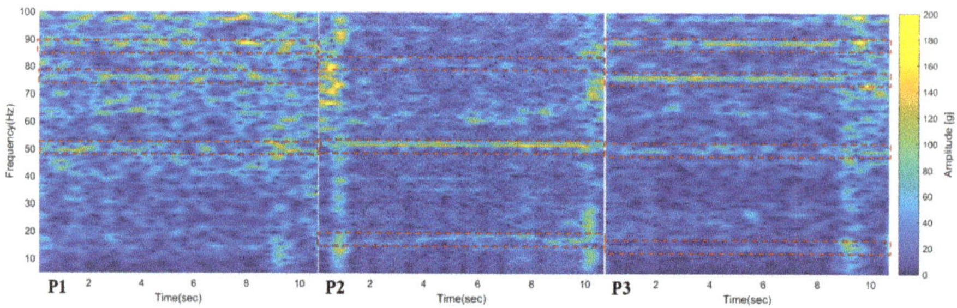

Figure 14. Short-time fourier transform (STFT) of vibration signals of P1–P3.

As for the milled surface, an obvious tool recession appears at P3, as shown in the red frame in Figure 15, while the situations are better in at P1 and P2. In addition, the quality of milled surfaces at P3 is the poorest compared with P1 and P2, and the vibration phenomenon on milled surface is getting severer from P1 to P3, indicating that configuration with higher first order frequency may lead to better milling performance, as mentioned in [20]. Whether there exits practical relevance between the value of natural frequency and machining performance can be a further research topic to develop more application of natural frequency prediction.

Figure 15. Milled surfaces of P1–P3.

5. Conclusions

It is impossible to measure all the natural frequency information of the whole robot workspace, but the information is vital in improving robot milling performance. In view of that, a method to predict the natural frequency is proposed in this paper. The core content of the method proposed is to construct a prediction model with sample configurations, which takes joint angles as input variables. Two insignificant variables q_1 and q_6 are abandoned and workspace is limited for practicability. Considering the nonlinear influence of joint angles on natural frequency, CCD method is used to select sample configurations from limited variable range, and EMT is applied to acquire the sample data of the example robot. Several models are fitted through the sample data. The model validation procedure proves that the model fitted and optimized by PLS method has the best prediction ability. Then the method proposed is applied completely onto another robot for universality verification, and the prediction performance turns out to be outstanding. Thus, this method can be applied on any 6R machining industrial robot to evaluate its natural frequency distribution, which can be used for in machining performance improvement.

Author Contributions: Conceptualization, J.S., W.Z. and X.D.; methodology, J.S. and X.D.; validation, J.S.; formal analysis, J.S.; investigation, J.S.; resources, W.Z.; data curation, J.S.; writing—original draft preparation, J.S.; writing—review and editing, J.S. and W.Z.; visualization, J.S.; supervision, D.X and W.Z.; project administration, W.Z.; funding acquisition, W.Z. All authors have read and agreed to the published version of the manuscript.

Funding: This research is supported by the National key R&D project—The Construction, Reference Implementation, and Verification Platform of Reconfigurable Intelligent Production System, China (Grant No. 2017YFE0101400).

References

1. International Federation of Robotics. Available online: https://www.ifr.org/# (accessed on 1 October 2020).
2. Chen, Y.; Dong, F. Robot machining: Recent development and future research issues. *Int. J. Adv. Manuf. Technol.* **2013**, *66*, 1489–1497. [CrossRef]
3. Kazerooni, H.; Bausch, J.J.; Kramer, B.M. An Approach to Automated Deburring by Robot Manipulators. *J. Dyn. Syst. Meas. Control.* **1986**, *108*, 354–359. [CrossRef]
4. Vergeest, J.S.; Tangelder, J.W. Robot machines rapid prototype. *Ind. Robot. Int. J.* **1996**, *23*, 17–20. [CrossRef]
5. Zieliński, C.; Mianowski, K.; Nazarczuk, K.; Szynkiewicz, W. A prototype robot for polishing and milling large objects. *Ind. Robot. Int. J.* **2003**, *30*, 67–76. [CrossRef]
6. Sun, L.; Liang, F.; Fang, L. Design and performance analysis of an industrial robot arm for robotic drilling process. *Ind. Robot. Int. J.* **2019**, *46*, 7–16. [CrossRef]
7. Guo, Y.; Dong, H.; Wang, G.; Ke, Y. A robotic boring system for intersection holes in aircraft assembly. *Ind. Robot. Int. J.* **2018**, *45*, 328–336. [CrossRef]
8. Matsuoka, S.-I.; Shimizu, K.; Yamazaki, N.; Oki, Y. High-speed end milling of an articulated robot and its characteristics. *J. Mater. Process. Technol.* **1999**, *95*, 83–89. [CrossRef]

9. Brunete, A.; Gambao, E.; Koskinen, J.; Heikkilä, T.; Kaldestad, K.B.; Tyapin, I.; Hovland, G.; Surdilovic, D.; Hernando, M.; Bottero, A.; et al. Hard material small-batch industrial machining robot. *Robot. Comput. Integr. Manuf.* **2018**, *54*, 185–199. [CrossRef]

10. Dong, X.; Zhang, W.; Sun, J. The estimation of cutting force coefficients in milling of thin-walled parts using cutter with different tooth radii. *Proc. Inst. Mech. Eng. Part B J. Eng. Manuf.* **2014**, *230*, 194–199. [CrossRef]

11. Dong, X.; Zhang, W. Chatter suppression analysis in milling process with variable spindle speed based on the reconstructed semi-discretization method. *Int. J. Adv. Manuf. Technol.* **2019**, *105*, 2021–2037. [CrossRef]

12. Dong, X.; Qiu, Z. Stability analysis in milling process based on updated numerical integration method. *Mech. Syst. Signal Process.* **2020**, *137*, 106435. [CrossRef]

13. Pan, Z.; Zhang, H.; Zhu, Z.; Wang, J. Chatter analysis of robotic machining process. *J. Mater. Process. Technol.* **2006**, *173*, 301–309. [CrossRef]

14. Guo, Y.; Dong, H.; Ke, Y. Stiffness-oriented posture optimization in robotic machining applications. *Robot. Comput. Integr. Manuf.* **2015**, *35*, 69–76. [CrossRef]

15. Lin, Y.; Zhao, H.; Ding, H. Spindle configuration analysis and optimization considering the deformation in robotic machining applications. *Robot. Comput. Integr. Manuf.* **2018**, *54*, 83–95. [CrossRef]

16. Chen, C.; Peng, F.; Yan, R.; Li, Y.; Wei, D.; Fan, Z.; Tang, X.; Zhu, Z. Stiffness performance index based posture and feed orientation optimization in robotic milling process. *Robot. Comput. Integr. Manuf.* **2019**, *55*, 29–40. [CrossRef]

17. Xiong, G.; Ding, Y.; Zhu, L.-M. Stiffness-based pose optimization of an industrial robot for five-axis milling. *Robot. Comput. Integr. Manuf.* **2019**, *55*, 19–28. [CrossRef]

18. Zaeh, M.F.; Roesch, O. Improvement of the Static and Dynamic Behavior of a Milling Robot. Mini Special Issue on Virtual Manufacturing. *Int. J. Autom. Technol.* **2015**, *9*, 129–133. [CrossRef]

19. Mejri, S.; Gagnol, V.; Le, T.-P.; Sabourin, L.; Ray, P.; Paultre, P. Dynamic characterization of machining robot and stability analysis. *Int. J. Adv. Manuf. Technol.* **2015**, *82*, 351–359. [CrossRef]

20. Bisu, C.; Cherif, M.; Gerard, A.; K'Nevez, J.Y. Dynamic Behavior Analysis for a Six Axis Industrial Machining Robot. *Adv. Mater. Res.* **2011**, *423*, 65–76. [CrossRef]

21. Mousavi, S.; Gagnol, V.; Bouzgarrou, B.C.; Ray, P. Model-based stability prediction of a machining robot. In *New Advances in Mechanisms, Mechanical Transmissions and Robotics*; Springer: Cham, Switzerland, 2017; pp. 379–387.

22. Karim, A.; Hitzer, J.; Lechler, A.; Verl, A. Analysis of the dynamic behavior of a six-axis industrial robot within the entire workspace in respect of machining tasks. In Proceedings of the 2017 IEEE International Conference on Advanced Intelligent Mechatronics (AIM), Munich, Germany, 3–7 July 2017; pp. 670–675.

23. Glogowski, P.; Rieger, M.; Bin Sun, J.; Kuhlenkötter, B. Natural Frequency Analysis in the Workspace of a Six-Axis Industrial Robot Using Design of Experiments. *Adv. Mater. Res.* **2016**, *1140*, 345–352. [CrossRef]

24. Guo, Y.; Dong, H.; Wang, G.; Ke, Y. Vibration analysis and suppression in robotic boring process. *Int. J. Mach. Tools Manuf.* **2016**, *101*, 102–110. [CrossRef]

25. Cordes, M.; Hintze, W.; Altintas, Y. Chatter stability in robotic milling. *Robot. Comput. Integr. Manuf.* **2019**, *55*, 11–18. [CrossRef]

26. Simões, J.A.; Coole, T.; Cheshire, D.; Pires, A.M. Analysis of multi-axis milling in an anthropomorphic robot, using the design of experiments methodology. *J. Mater. Process. Technol.* **2003**, *135*, 235–241. [CrossRef]

27. Wang, H.; Wu, Z.; Meng, J. *Partial Least-Square Regression—Linear and Nonlinear Methods*; National Defense Industry Press: Beijing, China, 2006; pp. 97–123.

Article

Semi-Active Magnetorheological Damper Device for Chatter Mitigation during Milling of Thin-Floor Components

Santiago Daniel Puma-Araujo [1], Daniel Olvera-Trejo [1,*], Oscar Martínez-Romero [1,*], Gorka Urbikain [2], Alex Elías-Zúñiga [1] and Luis Norberto López de Lacalle [2]

[1] Tecnologico de Monterrey, School of Engineering and Sciences, Ave. Eugenio Garza Sada 2501, Monterrey, N.L. 64849, Mexico; A00808090@itesm.mx (S.D.P.-A.); aelias@tec.mx (A.E.-Z.)

[2] Department of Mechanical Engineering, University of the Basque Country, Alameda de Urquijo s/n, 48013 Bilbao, Bizkaia, Spain; gorka.urbikain@ehu.es (G.U.); norberto.lzlacalle@ehu.eus (L.N.L.d.L.)

* Correspondence: daniel.olvera.trejo@tec.mx (D.O.-T.); oscar.martinez@tec.mx (O.M.-R.); Tel.: +52-(81)-8358-2000 (ext. 5430) (D.O.-T.)

Received: 2 July 2020; Accepted: 29 July 2020; Published: 31 July 2020

Abstract: The productivity during the machining of thin-floor components is limited due to unstable vibrations, which lead to poor surface quality and part rejection at the last stage of the manufacturing process. In this article, a semi-active magnetorheological damper device is designed in order to suppress chatter conditions during the milling operations of thin-floor components. To validate the performance of the magnetorheological (MR) damper device, a 1 degree of freedom experimental setup was designed to mimic the machining of thin-floor components and then, the stability boundaries were computed using the Enhance Multistage Homotopy Perturbation Method (EMHPM) together with a novel cutting force model in which the bull-nose end mill is discretized in disks. It was found that the predicted EMHPM stability lobes of the cantilever beam closely follow experimental data. The end of the paper shows that the usage of the MR damper device modifies the stability boundaries with a productivity increase by a factor of at least 3.

Keywords: thin-floor machining; chatter; magnetorheological damper; bull-nose end mill

---

1. Introduction

The manufacturing of parts such as impellers, ribs, blisks, and turbine blades among other components for the aerospace industry requires the machining of thin-wall and thin-floor features [1]. Titanium, aluminum alloys, and superalloys such as Inconel are used for these applications because of their good corrosion resistance, lightweight, and mechanical properties. Most of these components are machined as monolithic parts to improve their performance and their weight-resistant ratio. However, the lightweight design of these components has as a consequence thin walls and thin floors to achieve a good buy-to-fly ratio [2]. The low stiffness at specific part locations makes machining prone to unstable regenerative vibrations (chatter), which is detrimental during finishing machining operations, causing negative effects on surface quality and tool life. To avoid vibration, chip load and cutting forces are kept in conservative values, resulting in decreasing productivity. Therefore, the machining of monolithic parts represents a great technological challenge that is faced with several off-line and in-line solutions.

Literature offers in-line solutions from fixing perspectives [3], workpiece holders [4], stiffening devices [5], and active materials [6]. In fact, the use of these materials such as piezoelectric, magnetostrictive, magnetorheological (MR), and electrorheological (RT) materials has been implemented in actuators and sensors devices in manufacturing applications.

Among the active materials, also called smart materials, the MR and RT materials whose rheological behavior can be controlled externally through a magnetic and electrical field are mainly used as damper and spring devices. The MR fluids have a vast range of applications in different areas such as the automotive industry and civil engineering. For instance, the automotive industry has developed semi-active damper devices for automotive suspension, together with the control of the systems [7]. In civil engineering, MR damper devices are used to change the structural behavior of buildings during earthquakes [8]. In addition, the use of MR fluids is extended in several areas of engineering such as the research of vibrations of pipelines reported by Hui et al. [9]. Concisely, MR technology supports many applications over different industrial contexts.

In machining, it is a common practice to instrument MR and RT materials in the clamping systems so that workpiece properties can be modified by adjusting their damping and stiffness [10]. The MR and RT fluids can work at a wide frequency range; their response occurs in a short time and offers flexibility to be adapted to different machining processes. Recently, in [11], Fleischer et al. studied some industrial applications of this technology; however, the use of these smart materials in machining processes has evolved because of their feasibility of being implemented for scale-up production.

Segalman and Redmond [12] used ER fluids for chatter suppression; they applied a cyclic electrical field to the spindle quill, changing the natural frequency of the spindle, which disrupts the modulation of tool vibrations. They found significant reductions in vibration amplitudes through simulation of the milling process. Wang and Fei [13] used an ER fluid-filled sleeve to increase the stiffness and damping of a boring bar. They concluded that the applied electric field strength should be adjusted on-line according to the detected vibration signals. Mei et al. [14] used MR fluids to mitigate chatter on a boring bar by adjusting the bar stiffness via a magnetic field intensity. Similarly, Çeşmesi and Engin [15] developed an MR damper device that was implemented in a conventional impact machine. Their mathematical model was able to predict the behavior of the damper device by controlling the magnetic field. According to Som et al. [16], the attenuation rate of the semi-active MR actuator in a boring process can be up to 30% in wide frequency ranges of 400 to 600 Hz. Furthermore, MR devices have the capability of substantially improving surface finish and reducing tool wear in hard-turning operations [17]. The design of the MR damper device can be assisted by Finite Element Method simulation because it can identify the magnetic field magnitudes needed when located in a tool-holder [18]. It is interesting to point out that most implementations of MR damper devices in machining operations were mainly oriented to boring or turning operations. This could be explained since the MR damper devices could be easily attached to the tool-holder in lathe machines. However, its implementation in milling operations is very infrequent, but there have been a few attempts to apply MR technology to increase the damping of the workpiece. For instance, Ma et al. [19] demonstrated stability improvement in the thin-wall milling process.

The most common off-line approach is focused on the calculation of chatter-free cutting parameters based on the study of dynamics stability [20]. Special milling tools such as spherical or bull-nose tools are preferred for thin-floor machining in order to obtain a certain curvature between floors and walls; however, the geometry of the cutting edge has a strong influence over the forces in the tool axis direction, which typically coincides with the direction of the low-dynamic stiffness of thin floors. Since thin-floor dynamics are inherently complex due to the modal parameter variation that depends on the tool-workpiece location, three-dimensional stability diagrams for bull-nose milling tool cutting operations were calculated for identification of the location at which the spindle speed of the cutting tool is a chatter-free tuning interval value [21]. Of course, the accuracy of the location of stability zones depends on predicting the magnitude values of cutting forces. In this sense, Budak et al. [22] developed a method to predict the cutting force of any cutter geometry from orthogonal cutting data. The accuracy of the predicted forces was assessed with collected experimental data. The same method was also applied to ball-end milling tools [23]. By fitting pressure and friction coefficients on the rake and flank contact surfaces of ball cutters, Yucesan et al. [24] predicted with good accuracy the acting cutting forces in Cartesian directions.

Altintas [25] described the chip regeneration mechanism in three directions with a bull nose. The stability lobes were predicted in the frequency domain and verified experimentally. Other methods to predict stability are based on the time-domain discretization. For instance, Olvera et al. [26] proposed the Enhanced Multistage Homotopy Perturbation Method to compute stability lobes in one degree of freedom mimicking thin wall machining. Later, Urbikain et al. [27] proposed stability colormaps based on time-domain simulations of peripheral milling in thin-walled parts with barrel cutters. Regarding tool path planning in thin floors, Smith and Dvorak [28] proposed a tool path strategy that provided extra stiffness by leftover material in the corners; however, this strategy is limited for large areas consisting of mostly thin floors.

To implement reliable solutions for the machining operations of thin-floor components, two aspects need to be considered. Firstly, high accuracy is required for the determination of the modal parameters and the associated acting cutting forces for effective and reliable stability prediction. Secondly, it is important to acknowledge the shortcomings of active or semi-active damping techniques, including excessive setup times and customized designs that are only suitable for some specific machine processes. The use of MR fluids still is not mature in cutting operations and requires more investigation if one wants to enhance its industrial application.

Therefore, the aim of this work focuses on investigating how a semi-active MR damping device can influence the dynamics of cutting operations in thin-floor components. First, a stability prediction model is developed, and its accuracy is assessed by considering the milling operations of a thin-floor component with bull-end mills. Second, a semi-active damping device based on MR fluid is designed for improving chatter-free cutting conditions to increase productivity. In this sense, the proposed damper device is studied experimentally in order to improve the stability lobes for thin-floor machining operations.

This paper is summarized as follows. Section 2 focuses on developing the cutting force model for a bull-nose milling tool based on variable cutting coefficients. Section 3 describes the method used to compute the stability analysis in thin-floor machining and comparing theoretical prediction with experimental data. Section 4 describes the development of a semi-active MR damper device that is an influence on thin-floor dynamics to increase stability lobes and productivity. Finally, some conclusions are drawn.

2. A Mechanistic Model for Bull-Nose End Mills

The cutting forces in a flat end milling are dominant in the $x-$ and $y-$ directions. However, for ball and bull-nose end milling, the cutting forces have significant components in all three Cartesian directions [25,29,30]. For chatter analysis of thin floors using bull-nose end milling, the accuracy of the cutting force models is critical to predict reliable stability lobes. The semi-empirical models are based on the assumption that the cutting forces and the undeformed chip section are linked through the cutting coefficients. The shear coefficient represents the amount of force required for a tool to remove an uncut chip, while the edge coefficient is related to the friction forces.

A bull-nose end milling tool with a tool diameter D, helix angle β, and tooth N_z, is defined by corner radius R (also called the ball radius). The bull-nose end mill used was purchased from 3G tool supplier and the main geometry characteristics described earlier were measured with an Alicona microscope model InfiniteFocusG5 (Alicona Imaging GmbH, Pforzheim, Germany) together with a 2.5X objective. As illustrated in Figure 1, optical microscope photographs confirm a tool diameter of 16 mm and a nose radius of 2.5 mm.

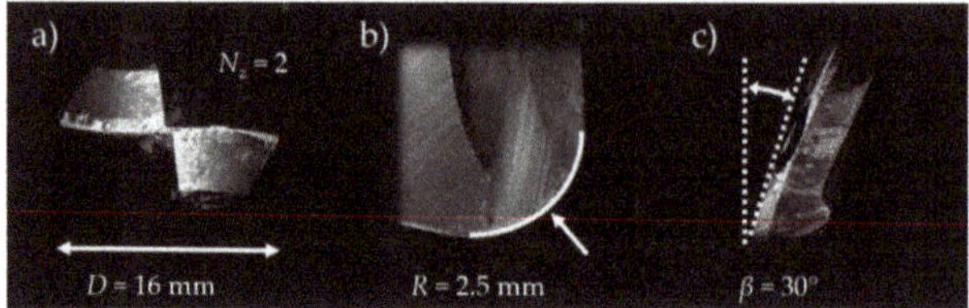

Figure 1. Composition of optical microscope photographs of the bull-nose end mill taken with an Alicona microscope: (**a**) bottom view (diameter measurement); (**b**) frontal view (nose radius measurement); (**c**) lateral view (helix angle measurement).

2.1. Cutting Force Model

A force model was developed to obtain an instantaneous cutting force magnitude for a given depth of cut. It is based on a mechanistic approach that assumes a relationship between forces and the uncut chip thickness by means of the cutting coefficients. Cutting models have been developed for ball-end mills that present similarities with the geometry of a bull-nose end mill. For instance, Yucesan et al. [24] proposed a model based on an analytic representation of ball-shaped helical flute geometry. Budak et al. reported the prediction of forces from the orthogonal cutting data [22] and for better detail, an oblique cut-off model presented by Altintas and Lee [31] was evaluated in sections. However, the development of a generalized and accurate model for cutting coefficients for a bull-nose end mill has not been set. Campa et al. [21] developed a model using averaged cutting coefficients in the toroid part, but they recognized that the flank of the tool presented a different behavior.

The cutting force model is established by introducing two coefficients for the tangential, radial, and axial directions: one is associated with cutting (shearing) K_c, which is directly related to the dynamic chip thickness $h(\phi, \kappa)$ and the second is associated with the friction (rubbing) coefficient K_e. As shown in Figure 2, the differential forces components dF_t, dF_r, and dF_a are orientated for each differential cutting edge according to the cutting edge angle $\kappa(z)$. The cutting edge of the tool is discretized in disks with a thickness of $b = 1$ mm along the z-direction. With this discretized approach, it is possible to understand in detail the behavior of the cutting coefficients along the cutting edge in the toroid part. Figure 2 illustrates the cutting parameters of thin-floor machining: axial depth of cut a_p along the z-direction, radial immersion a_e of cut in the y-direction, and spindle speed n.

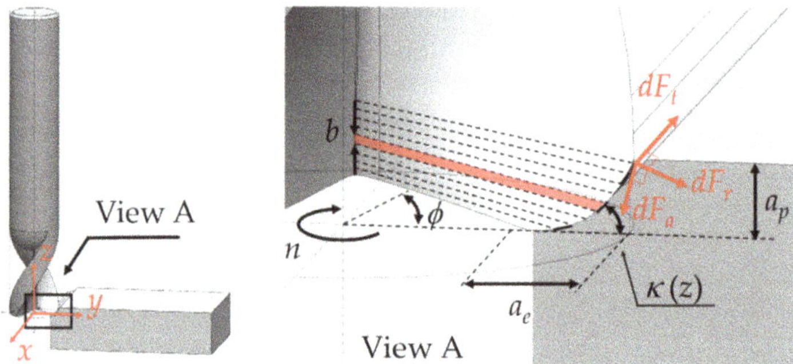

Figure 2. Main cutting parameters in milling operation with a bull-nose end mill: axial discretization and differential cutting force components.

The mechanism of the bull-nose end mill is modeled using a mechanistic linear model. Here, the differential cutting forces on the local tool system follow the analytical approach in [32]:

$$dF_t\!\left(\phi_j\right) = K_{tc}h\!\left(\phi_j,\kappa\right)dz + K_{te}dz$$
$$dF_r\!\left(\phi_j\right) = K_{rc}h\!\left(\phi_j,\kappa\right)dz + K_{re}dz \tag{1}$$
$$dF_a\!\left(\phi_j\right) = K_{ac}h\!\left(\phi_j,\kappa\right)dz + K_{ae}dz$$

where $dF_{t,r,a}$ are the differential force components, $\phi_j(t)$ is the angle of each tooth on the cutter, which change over time, and $h\!\left(\phi_j,\kappa\right)$ is the instantaneous chip thickness as described by Compean et al. in [33]:

$$h\!\left(\phi_j,\kappa\right) = f_z \sin\phi_j \sin\kappa. \tag{2}$$

Here, f_z is the feed rate per tooth and κ is the cutting-edge angle for a given axial slice. By means of a rotation matrix, it is possible to transform the differential cutting forces to Cartesian coordinates

$$\left\{\begin{array}{c} dF_x\!\left(\phi_j,z\right) \\ dF_y\!\left(\phi_j,z\right) \\ dF_z\!\left(\phi_j,z\right) \end{array}\right\} = \left[\begin{array}{ccc} \cos\phi_j & \sin\phi_j\sin\kappa & -\sin\phi_j\cos\kappa \\ \sin\phi_j & -\cos\phi_j\sin\kappa & \cos\phi_j\cos\kappa \\ 0 & -\cos\kappa & -\sin\kappa \end{array}\right]\left\{\begin{array}{c} dF_t\!\left(\phi_j,z\right) \\ dF_r\!\left(\phi_j,z\right) \\ dF_a\!\left(\phi_j,z\right) \end{array}\right\}. \tag{3}$$

2.2. Characterization Procedure

The characterization procedure assumes the linear relationship between the averaged cutting forces $\tilde{F}$ and the feed rate f_z. This relationship is established as follows:

$$\tilde{F} = f_z\cdot\tilde{F}_c + \tilde{F}_e. \tag{4}$$

Here, $\tilde{F}_c$ and $\tilde{F}_e$ are the cutting shear and edge components, respectively. Developing Equation (4) in terms of differential cutting forces, the following expression is obtained:

$$F(\phi) = \int_{z_1}^{z_2} dF(\phi,z) = \int_{z_1}^{z_2} dF_c(\phi,z)\cdot f_z + \int_{z_1}^{z_2} dF_e(\phi,z). \tag{5}$$

To determine the average cutting force per tooth, Equation (5) is integrated within the limits of the entrance (ϕ_s) and exit (ϕ_e) angles, and it is divided by the pitch angle $2\pi/N_z$:

$$\tilde{F} = \frac{N_z}{2\pi}\int_{\phi_s}^{\phi_e} F d\phi. \tag{6}$$

Therefore, the Cartesian components of the average forces for the tool are:

$$\tilde{F}_{xc} = \frac{N_t}{2\pi}\int_{\phi_s}^{\phi_e}\int_{z_1}^{z_2}\left(K_{tc}\sin\phi\cos\phi\sin\kappa + K_{rc}\sin^2\phi\sin^2\kappa - K_{ac}\sin^2\phi\sin\kappa\cos\kappa\right)dzd\phi$$
$$\tilde{F}_{yc} = \frac{N_t}{2\pi}\int_{\phi_s}^{\phi_e}\int_{z_1}^{z_2}\left(K_{tc}\sin^2\phi\sin\kappa - K_{rc}\sin\phi\cos\phi\sin^2\kappa + K_{ac}\sin\phi\cos\phi\sin\kappa\cos\kappa\right)dzd\phi \tag{7}$$
$$\tilde{F}_{zc} = \frac{N_t}{2\pi}\int_{\phi_s}^{\phi_e}\int_{z_1}^{z_2}\left(-K_{rc}\sin\phi\sin\kappa\cos\kappa - K_{ac}\sin^2\phi\sin\kappa\right)dzd\phi.$$

The final step consists of computing shear and edge coefficients by solving the system of Equation (7).

2.3. Experimental Procedure

A total of 120 cuttings were performed for 100% radial immersion in aluminum 7075T6 during dry machining. The forces were recorded by using a dynamometer 9257B (Kistler Group, Winterthur, Switzerland) for the set of cutting conditions listed in Table 1. The spindle speed is set at 3000 rpm based on the dynamometer's natural frequency to avoid the amplification of milling forces. The force signals were acquired using a VibSoft-20 acquisition card (Polytec GmbH, Waldbronn, Germany) at a sample rate of 48 kHz and processed in a custom-made MATLAB app to remove drift and noise phenomena.

Cutting forces data were collected for increments of 0.1 mm in the axial depth of cut with the aim of minimizing the model errors and obtaining detailed information of the cutting coefficients along the cutting edge in the toroid section. For a given axial depth of cut, four values of feed per tooth were tested as listed in Table 1.

Table 1. Characterization test parameters.

Spindle speed	3000 rpm
Radial Immersion	16 mm, down milling
Depth of cut $\left(a_p\right)$	0.1–3.0 [mm]
Feed per tooth (f_z)	0.05, 0.10, 0.15, 0.20 [mm/tooth]

For a set of experiments with a specific axial depth of cut, the system of Equations (7) is solved to obtain the K_c and K_e cutting coefficients. The computed data were fitted by using a second-order polynomial function as listed in Table 2.

Table 2. Cutting coefficients of the tool.

Cutting Coefficients	Toroid [N/mm^2]	Flank [N/mm^2]
K_{tc}	$138.89z^2 - 625z + 1136.9$	434
K_{rc}	$132.98z^2 - 581.96z + 785.4$	149.3
K_{ac}	$37.57z^2 - 159.06z + 171.06$	2.8

Figure 3 displays the fitted model for shear coefficients in the tool coordinate system (tangential, radial, and axial directions). It can be seen from Figure 3 that the cutting coefficients, in the range of 0.1–1.5 mm, change significantly whereas for the section approaching the flank section of the cutting edge, these remain almost constant. This can be explained from Figure 3d, in which the cutting-edge angle starts from zero at the bottom of the cutting edge and increases to 90° at the flank tool section. In a square end mill, where the cutting-edge angle is 90° at any edge location, the cutting coefficients remain constant. However, for the bull-nose end mill, the cutting coefficient exhibits nonlinear behavior. For this reason, the proposed mechanistic model is important to predict accurately stability boundaries.

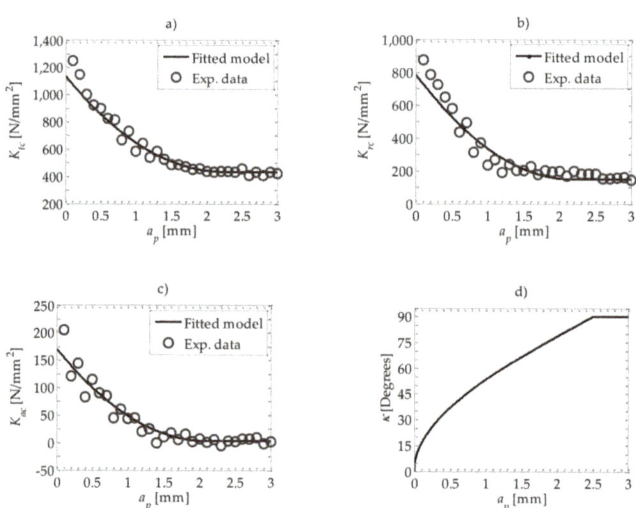

Figure 3. Experimental cutting coefficients (circle marks) vs. fitted model (solid lines) along the toroidal section: (**a**) tangential; (**b**) radial; (**c**) axial directions; (**d**) κ angle along the toroid section.

Notice from Figure 3a–c that the cutting coefficients in all directions present high values for the toroid section: almost three times higher in the radial direction (see Figure 3b), and almost five times higher in the axial direction (see Figure 3c) compared to the flank section of the cutting edge, which could affect significantly the critical depth of cut of thin-floors machining.

In order to validate the proposed model, slot milling tests were performed. Figure 4 shows the comparison between the proposed cutting force-fitted model and experimental forces for two sets of cutting conditions. The accuracy of the proposed model to fit experimental data is shown in Figure 4; it is evident that in both cases, the proposed forced model predicts the experimental data well.

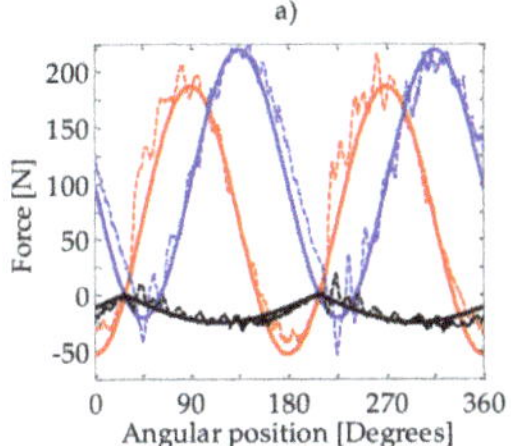
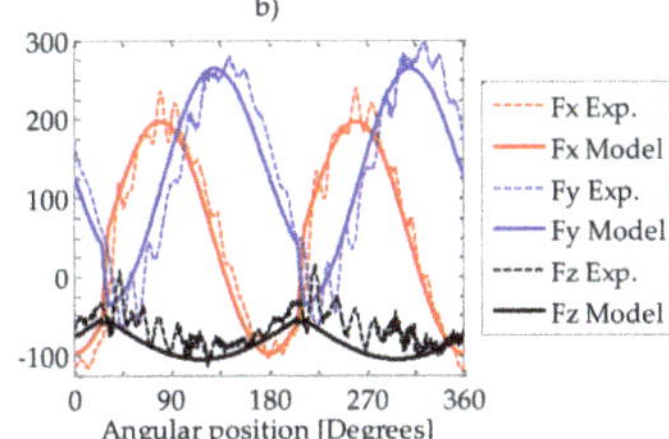

Figure 4. Comparison between experiment data and the proposed force model at 3000 rev/min and the cutting conditions: (**a**) $a_p = 0.5$ mm, $f_z = 0.2$ mm/tooth; (**b**) $a_p = 1.5$ mm, $f_z = 0.15$ mm/tooth.

3. Stability Analysis and Experimental Validation

3.1. Stability Analysis for Thin-Floor Machining

In finish milling operations of thin-floor components where the dominant mode is oriented in the z-direction, the stability model can be simplified considering only the Frequency Response Function (FRF) of the thin floor, so the system becomes a one degree of freedom (1 dof) model as described by Campa et al. [30]. In this case, the dynamic chip thickness is determined from:

$$h\big(\phi_j(t)\big) = \cos \kappa \Delta z \tag{8}$$

where $\phi_j(t)$ is the angular position of the cutting edge j and $\Delta z = (z(t) - z(t - \tau))$. The equation of motion that describes the dynamics of the thin floor is given as:

$$\ddot{z}(t) + 2\zeta\omega_n\dot{z}(t) + \omega_n^2 z(t) = \frac{F_z(t)}{m_m}(z(t) - z(t - \tau)) \tag{9}$$

where m_m is the modal mass, ζ describes the damping ratio, ω_n is the angular natural frequency, and F_z is the cutting force in the z-direction, which is given as:

$$F_z(t) = \sum_{j=0}^{N_t-1} g\big(\phi_j(t)\big)h\big(\phi_j(t)\big)\big(-K_{rc}\cos^2\kappa - K_{ac}\sin\kappa\cos\kappa\big). \tag{10}$$

Here, $g\big(\phi_j(t)\big)$ is a switching function that returns the value of one when the tooth cutter is cutting and the value of zero otherwise. The matrix representation of Equation (9) becomes:

$$\dot{\mathbf{z}}(t) = \mathbf{A}(t)\mathbf{z}(t) + \mathbf{B}(t)\mathbf{z}(t - \tau) \tag{11}$$

where $\mathbf{z} = \begin{bmatrix} z, \dot{z} \end{bmatrix}^T$, $\mathbf{A}(t + \tau) = \mathbf{A}(t)$, $\mathbf{B}(t + \tau) = \mathbf{B}(t)$, and τ is the time delay. Following the Enhanced Multistage Homotopy Perturbation Method (EMHPM) described in [34], Equation (11) is written in an equivalent form as:

$$\dot{\mathbf{z}}_i(T) - \mathbf{A}_t\mathbf{z}_i(T) \approx \mathbf{B}_t\mathbf{z}_i^\tau(T) \tag{12}$$

where $\mathbf{z}_i(T)$ indicates the m-order solution for Equation (11) that satisfies the initial conditions $\mathbf{z}_i(0) = \mathbf{z}_{i-1}$, $\mathbf{A}_t$ and $\mathbf{B}_t$ are the periodic matrix whose values vary with time t. The period $[t_0 - \tau, t_0]$ is discretized in N points to approximate the term $\mathbf{z}_i^\tau(T)$ to the delay in Equation (12). The function $\mathbf{z}_i^\tau(T)$ could be approximate as a first-polynomial representation:

$$\mathbf{z}_i^\tau(T) \approx \mathbf{z}_{i-N} + \frac{N-1}{\tau}(\mathbf{z}_{i-N+1} + \mathbf{z}_{i-N})T \tag{13}$$

Next, we define $\mathbf{z}_i \equiv \mathbf{z}_i(T_i)$ and then, Equation (12) is substituted into Equation (13) to get:

$$\dot{\mathbf{z}}_i = \mathbf{A}_t\mathbf{z}_i + \mathbf{B}_t\mathbf{z}_{i-N} - \frac{N-1}{\tau}\mathbf{B}_t\mathbf{z}_{i-N}T + \frac{N-1}{\tau}\mathbf{B}_t\mathbf{z}_{i-N+1}T \tag{14}$$

where

$$\mathbf{A}_t = \begin{bmatrix} 0 & 1 \\ -\omega_n^2 + \frac{F_z(t)}{m_m} & -2\zeta\omega_n \end{bmatrix}, \mathbf{B}_t = \begin{bmatrix} 0 & 0 \\ -\frac{F_z(t)}{m_m} & 0 \end{bmatrix}. \tag{15}$$

To compute the stability lobes of Equation (14), the procedure described in [26] is followed. Then, it is assumed that Equation (14) can be written as a function of discretized states:

$$\mathbf{z}_i(T) \approx \mathbf{P}_i(T)\mathbf{z}_{i-1} + \mathbf{Q}_i(T)\mathbf{z}_{i-N} + \mathbf{R}_i(T)\mathbf{z}_{i-N+1} \tag{16}$$

where

$$\mathbf{P}_i(T) = \sum_{k=0}^{m} \frac{1}{k!}\mathbf{A}_t^k T^k,$$

$$\mathbf{Q}_i(T) = \begin{cases} \sum_{k=1}^{m} \frac{N-1}{(k+1)!(\tau)}\mathbf{A}_t^{k-1}\mathbf{B}_t T^{k+1} & m \geq 1 \\ 0 & m = 0 \end{cases} \tag{17}$$

$$\mathbf{R}_i(T) = \begin{cases} \sum_{k=1}^{m} \frac{1}{k!}\mathbf{A}_t^{k-1}\mathbf{B}_t T^k - \mathbf{Q}_m & m \geq 1 \\ 0 & m = 0 \end{cases}.$$

Thus, the approximate solution of Equation (16) can be written as a discrete map $\mathbf{w}_i = \mathbf{D}_i\mathbf{w}_{i-1}$. By following the Floquet theory [35,36], the transition matrix $\mathbf{\Phi}$ of Equation (16) is calculated over period $\tau = (N-1)\Delta t$ by coupling each solution of $\mathbf{D}_i$, $i = 1, 2, \ldots N-1$ to get:

$$\mathbf{\Phi} = \mathbf{D}_{N-1}\mathbf{D}_{N-2}\ldots\mathbf{D}_2\mathbf{D}_1. \tag{18}$$

Finally, the stability lobes of Equation (11) are determined by computing the eigenvalues of the transition matrix. Notice that the 1 dof model proposed in Equation (9) assumes a single vibration mode; however, its application to a multi-degree of freedom system could be performed by computing the stability lobes for each mode.

3.2. Experimental Validation of Stability Lobes

Experimental milling cutting tests were performed in a Makino F3 machining center. A monolithic cantilever beam of $116 \times 177 \times 12.7$ mm was used to mimic a 1 dof thin-floor behavior in the z-direction defined as shown in Figure 5. A 7075T6 aluminum workpiece was positioned on the cantilevered plate using two screws to maintain its dynamic characteristics during several passes (total mass does not change substantially). In the thin-floor milling process, the cutting tool is much stiffer in comparison with the workpiece properties. For that reason, the dynamic response of a thin floor is the most important factor during machining. The workpiece FRF was measured with a compact laser vibrometer, model CLV-2534 (Polytec GmbH, Waldbronn, Germany) and a impulse force hammer, Model 9722A (Kistler Group, Winterthur, Switzerland). The fitted modal parameters listed in Table 3 were performed using CutPro Simulation Software (Manufacturing Automation Laboratories Inc., Vancouver, BC, Canada).

Table 3. Measured modal parameters of the workpiece.

Mode z	f [Hz]	k [N/m]	ξ
1	93	6.59E5	0.003
2	304	4.89E6	0.004

In order to validate the dynamics of the milling process together with the proposed cutting force model, stability lobes for 2.5 mm radial immersion were computed and shown in Figure 6 by using the EMHPM of Equation (11). As described in Section 2.2, the cutting forces model was considered for a range from 0 to 3 mm in the z-direction. The stability lobes verification was studied through the measurement of the vibrations of the workpiece mounted on the cantilevered plate. An accelerometer type 8778A500M14 (Kistler Group, Winterthur, Switzerland). weighing 0.4 g was used to acquire the vibrations, and a tachometer model FS-V31 (Keyence Corporation, Osaka, Japan) was used to detect every single revolution of the spindle as shown in Figure 5.

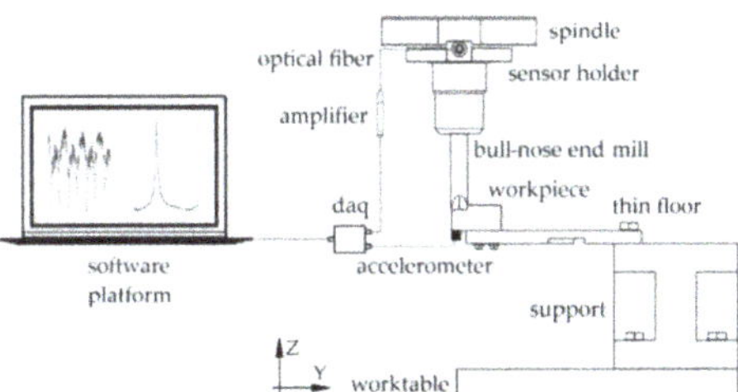

Figure 5. Experimental setup used for stability verification. An accelerometer is attached in the workpiece to detect vibrations, while the optical sensor is used to measure the spindle rotation.

The signal processing consisted of two steps. First, raw signals were filtered with a low-pass filter with a cut-off frequency of 5 kHz. Additionally, the tachometer signal was manipulated to include a virtual pulse in the middle of two consecutive detected pulses that allows us to sample the accelerometer signal according to the tool passing period (two flutes impact the workpiece during each revolution). Secondly, the analysis of the accelerometer signal was performed via wavelet transform (CWT), power spectral density (PSD), and Poincaré diagrams. By comparing experimental and theoretical chatter frequencies [37], all this information was used to identify and categorize every single cutting test as having stable or unstable cutting parameters. After completing the signal process and categorization, experimental data were superposed in Figure 6.

The stable cuts in Figure 6 are denoted by circles, the unstable quasi-periodic cuts are denoted by squares, and the unstable double periodic cuts are denoted by triangle symbols. Experimental results denoted by diamonds represent experimental conditions without dominant chatter frequencies, which makes the classification as either unstable or stable challenging.

From Figure 6, three cutting tests that describe typical stable/unstable behavior have been selected (points A, B, and C). Point A describes unstable period-doubling chatter, Point B describes stable cutting conditions, and Point C describes unstable double periodic. Since the CWT analysis allows us to observe how frequencies evolve in the time domain, then this analysis was applied to accelerometer data as shown in Figure 7a–c. To relate the detected frequency in relation to the excitation, the following normalized frequency is introduced:

$$f_n = f_w / f_{TPE}. \tag{19}$$

In Equation (19), f_w represents the wavelet transform frequency and f_{TPE} represents the tooth passing excitation frequency. The overlapped circles and cross white marks in Figure 7a–c identify normalized frequencies of $f_n = 1$ and $f_n = 0.5$, respectively. In Figure 7d,e, the acceleration sampling (red circles) denotes vibration amplitude when the cutting tool hits the workpiece. Poincare diagrams (Figure 7g–i) were constructed with the acceleration sampling data versus 20-delayed samples. The power spectral density (PSD) in Figure 7j–l describes the power density from the acceleration data in the frequency domain.

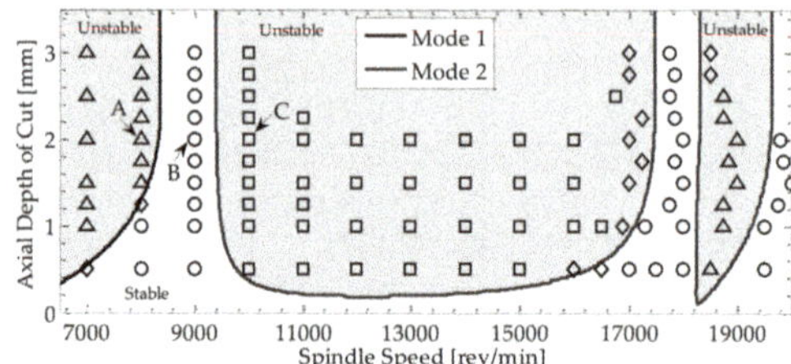

Figure 6. Predicted stability lobes obtained from Equation (11) together with experimental data in a down-milling operation. (○ stable cut, □ quasi-periodic chatter, Δ double periodic chatter, and ◊ border cutting parameters).

Point A (Figure 7a,d,g,j) exhibits unstable period-doubling chatter, meaning that each tooth passing the excitation period is half of the dominant period in the acceleration signal. Even the PSD plot (Figure 7j) exhibits several chatter frequencies, while the doming frequency almost coincides with the cross marks, which confirms the period-doubling chatter. Due to all these chatter frequencies, the Poincaré diagram in Figure 7g draws two semi-elliptical trajectories. This behavior contrasts with a pure period-double chatter experiment (data not shown) that draws two concentration of samples; in other words, the workpiece vibrates with a dominant frequency that is half that of the tool passing excitation frequency.

Point B (Figure 7b,e,h,k) shows stable cutting conditions. In this case, the dominant frequency in the CWT is related to the tool passing frequency (circle marks) during several revolutions. Notice as well that for each period τ, two stains appear corresponding to one vibration cycle: one peak and one valley. As a result, the peak frequencies in the PSD plot coincide with multiples of the tool passing frequency. A Poincaré diagram for stable cutting (Figure 7h) shows the concentration of sampled acceleration data, meaning the vibration amplitude is similar every time a flute hits the workpiece.

Point C describes Hopf instabilities; this instability draws in the PSD plot-dominant quasi-periodic chatter frequencies. Notice that the frequency intensity in the CWT (Figure 7c) is below normalized frequency $f_n = 1$, and those values change in time. In contrast with Point A and B, the PSD (Figure 7l) shows multiple chatter frequencies, which are responsible for several loops in the Poincaré diagram.

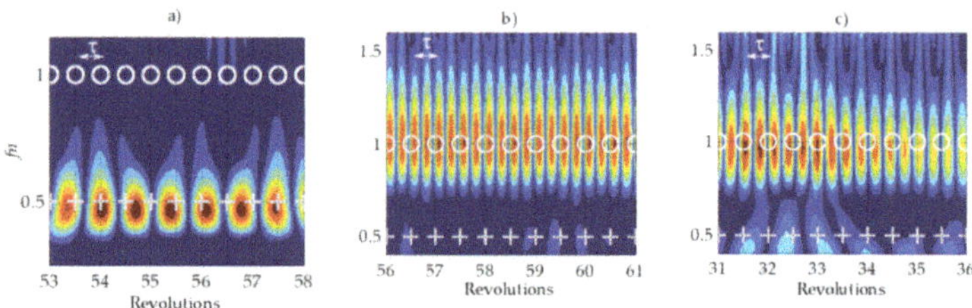

Figure 7. *Cont.*

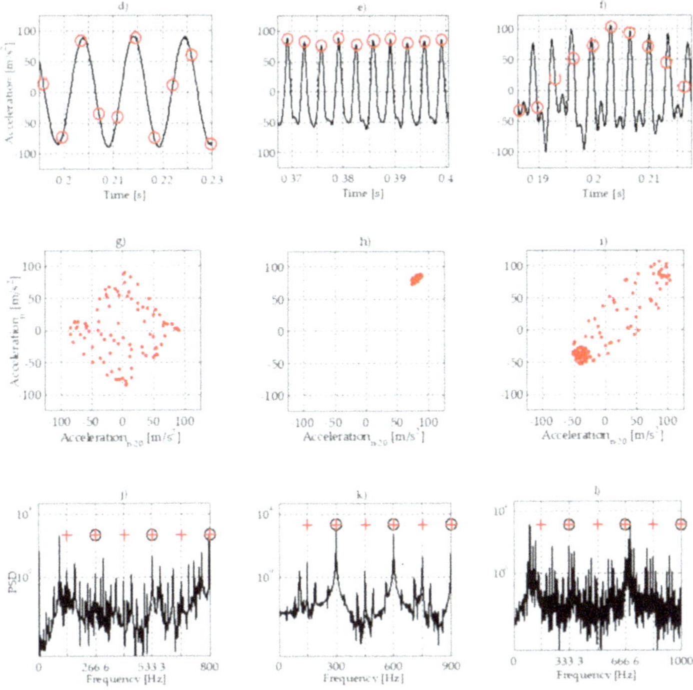

Figure 7. Continuous wavelet transforms (**a,b,c**), sampled acceleration signals (**d,e,f**), Poincaré diagrams (**g,h,i**), and power spectral density (**j,k,l**) for unstable cut A (**a,d,g,j**), stable cut B (**b,e,h,k**), and quasi-periodic unstable cut C (**c,f,i,l**).

4. Semi-Active Magnetorheological Damper Device for Chatter Mitigation

4.1. Experimental Setup of MR Damper Assembly

The experimental setup is implemented as shown in the optical photograph (Figure 8a). This system uses the same components described in Section 3.2, but now, it also includes the semi-active MR damper device. The MR damper device is a pool-like container or chamber fabricated in acrylic material; a non-magnetic material avoids reducing the magnetic flux density generated by an electromagnet. The acrylic chamber contains the magnetorheological fluid with an internal diameter of 200 mm and 19 mm height. In order to avoid spills, an acrylic plate is placed covering the area not in contact with the beam. The MR damper assembly is supported with 3D-printed columns so that an electromagnet could be placed between the worktable and the MR damper. An HP/Agilent power supply, model 6032A was connected to a electromagnet, model VEM200 (Electroimanes NAFSA, Gernika, Spain), which accepts a maximum input current of 3.4 A at 25 V. Once the electromagnet is energized independently of the machining, it generates a magnetic flux density that changes the yield stress of the MR fluid. The chamber was fully filled with an MRF-122EG MR fluid (LORD Corporation, Cary, NC, USA).

The cross-section A – A presented in Figure 8b shows a better view of the experimental setup and allows us to observe that the beam is sunk about 1 mm in the MR fluid with a contact area of about 50 cm^2. It is important to sink the beam to guarantee that it is always in contact with the MR fluid during machining, because the amplitude of the vibrations is expected to be in the range of ± 1 mm. Once the workpiece is attached to the beam, the experimental setup is ready for FRF measurements and experimental milling cuttings.

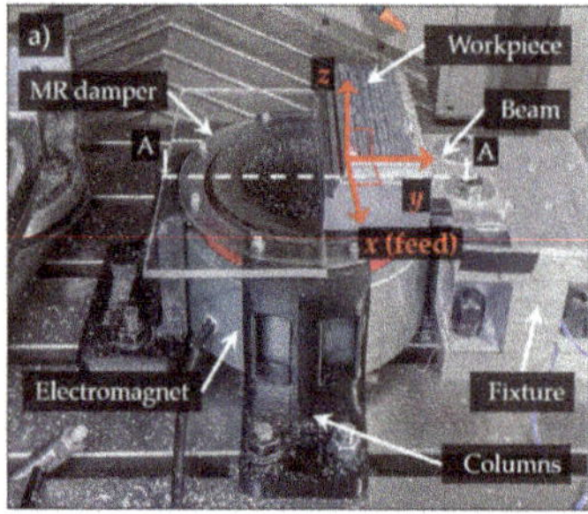
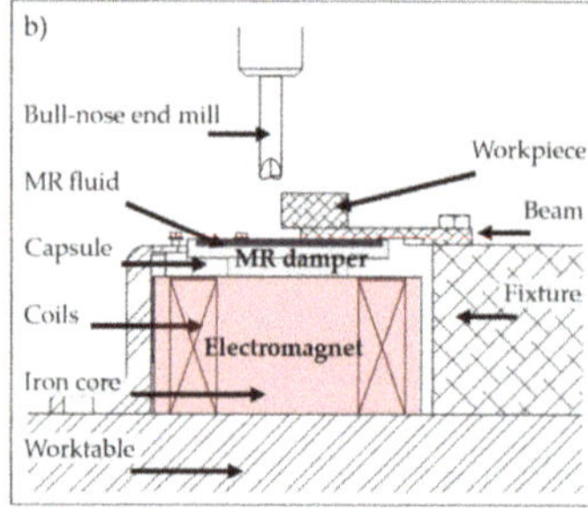

Figure 8. (**a**) Optical image of the experimental setup and magnetorheological (MR) damper installation; (**b**) Schematic of cross-section A–A, the assembling of capsule–MR fluid is placed above the electromagnet.

4.2. Modified FRF of the Cantilever Beam

For this experimental setup, to validate the effectiveness of the MR damper device, a new cantilever beam made of aluminum alloy 7075T6 was designated with dimensions of $103 \times 127 \times 12.7$ mm to induce specific modal parameters and study the stability boundaries of thin walls with a very low critical depth value (below 0.5 mm). In order to evaluate the effects of the MR damper over the stability boundaries, three levels of MR damper performance were studied: highest, 3.4 A; medium, 1.7 A; and lowest, off. The magnetic flux density produced by the electromagnet was measured experimentally with an F.W. BELL gaussmeter, model 5170. Figure 9 shows the magnetic flux density measured from the center of the cylindrical electromagnetic outwards along the radial direction. As expected, the magnitude of the magnetic flux density is almost constant in the radial direction, and the magnitude is proportional to the supplied electrical current. Measurements after 40 mm were not considered, since the cantilever beam was not in contact with the MR fluid.

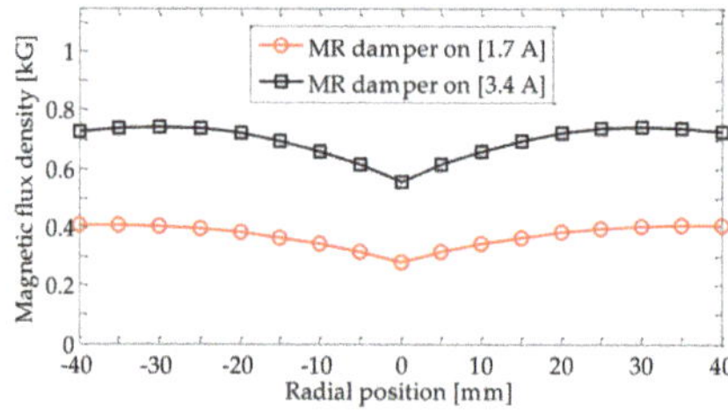

Figure 9. MR damper device magnetic flux density distribution.

The yield stress of the MR fluid changes depending on the magnetic flux density at which it is subjected. It is expected that the dynamic properties of the system imposed will be modified because of the effects of the MR damper device. The FRF was experimentally measured for 4 scenarios and plotted in Figure 10. The scenarios are (1) no damper, (2) damper off, (3) damper on with 1.7 A, and (4) damper on with 3.4 A. The technique used to acquire the FRF and the parameter fitting procedure are similar to those of Section 3.1. Table 4 summarizes the modal parameters computed for each scenario.

Notice from Table 4 that the natural frequency of the beam, without a damper, is 190 Hz and the damping ratio is 0.007. When MR fluid is in contact with the thin floor and off, the natural frequency shifts to 180 Hz, and the damping ratio increased by about 50%. Note that when the MR damper is off, the electromagnet is not active, and no electrical current is provided, but it is still in contact with the cantilever beam and the FRF is affected. Table 4 shows that if the MR damper is turned on and subjected to two electrical currents values, the damping ratio of the system increases proportionally.

The damping of the system depends on the yield stress of the magnetorheological fluid, which can be estimated by using the technical data sheet from the fluid manufacturer [38].

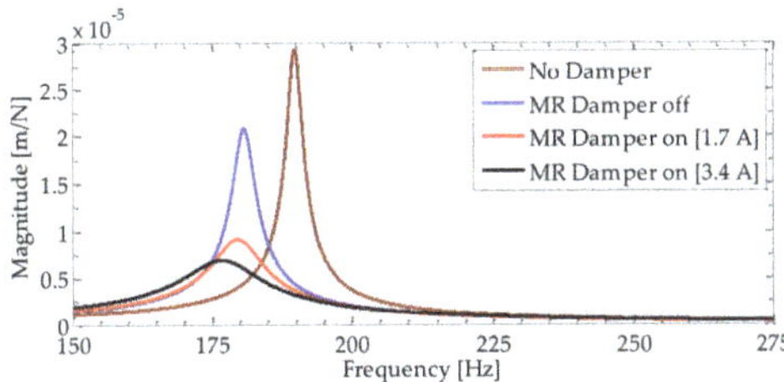

Figure 10. Measured Frequency Response Function (FRF) of the thin floor with the MR damper.

Table 4. Modal parameters of the cantilever beam under the action of a semi-active MR damper device.

Electrical Current [A]	Average Magnetic Flux Density [G]	Yield Stress [KPa]	Natural Frequency f [Hz]	Stiffness k [N/m]	Damping ξ
No Damper	-	-	190	2.40×10^6	0.007
Off–0	-	0	180	2.18×10^6	0.011
On–1.7	372	6	178	2.04×10^6	0.023
On–3.4	695	12	175	2.07×10^6	0.036

4.3. Experimental Determination of Stability Lobes under the Action of the MR Damper Device

Notice when the MR damper device was placed on the cantilever beam, the damping ratio of the beam change substantially, as listed in Table 4. Once the modal parameters were computed, the stability lobes were plotted in Figure 11 following the same methodology described in Section 3.1. Since the MR damper device increases the damping ratio in the system, the stability boundaries not only shift toward higher axial depth but also shrink, which substantially allows more productive cutting combinations. For instance, when the MR damper device is turned off and cuts are performed at 7000 rev/min, the critical axial depth of cut is close to 0.25 mm, but this value increases to 0.75 and to 1.4 mm when the current in the MR damper is adjusted at 1.7 and to 3.4 A, respectively. In order to validate the effect of the MR damper over the stability behavior, three cutting tests were selected, and these were named points D, E, F as illustrated in Figure 11. All of them correspond to spindle speeds of 7000 rpm but with different axial depth-of-cut values and subjected to specific scenarios. The accelerometer signal of point D was acquired when the MR damper was turned off, while the acceleration for points E and F were collected when the MR damper was turned on with 1.7 and 3.4 A, respectively.

Figure 12a shows the CWT obtained from the acceleration signal collected for point D (a_p = 0.5 mm). Notice that this CWT plot illustrates dominant frequencies (red colors) lower than the normalized f_n = 1, which indicates that cutting is an unstable quasi-periodic chatter test, as predicted by the stability lobes in Figure 11. This is also confirmed with the sampled acceleration signal plotted in Figure 12d, since the amplitudes of the samples are not similar. The same combination of cutting parameters (a_p = 0.5 mm and n = 7000 rev/min) are now analyzed when the MR damper device is turned on with 1.7 A. Notice from Figure 12b that for this cutting test (point E), the CWT describes dominant frequencies that are very close to normalized frequency f_n = 1, and thus, this is an indication of stable cutting conditions. This result agrees with the stability lobes plotted in Figure 11, since the location of point E indicates stable cutting conditions because it is below the critical axial depth of cut value of 0.75 mm. Similarly, point F (a_p = 1.25 mm) exhibits stable cutting conditions, since the critical depth of cut has now the value of 1.5 mm. It is important to observe that the acceleration amplitude for experimental test F (Figure 12f) is even lower than the value recorded for point E (Figure 12e), confirming that the MR damper device offers higher damping and is an effective technique to avoid unstable vibrations.

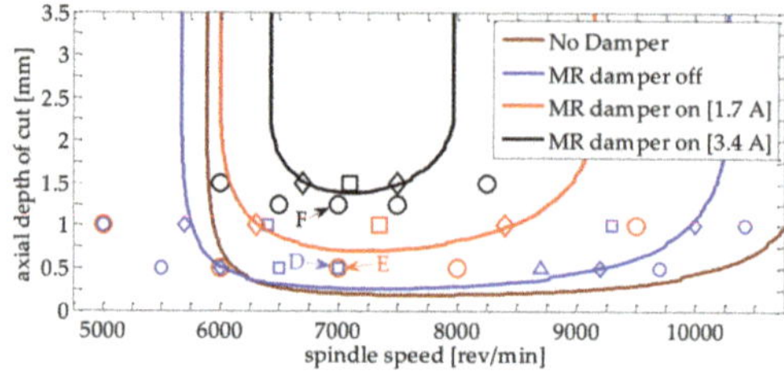

Figure 11. Stability lobes diagrams for three scenarios of magnetic flux density applied to the MR damper device together with experimental data in a down-milling operation. (○ stable cut, □ quasi-periodic chatter, △ double periodic chatter, and ◊ border cutting parameters).

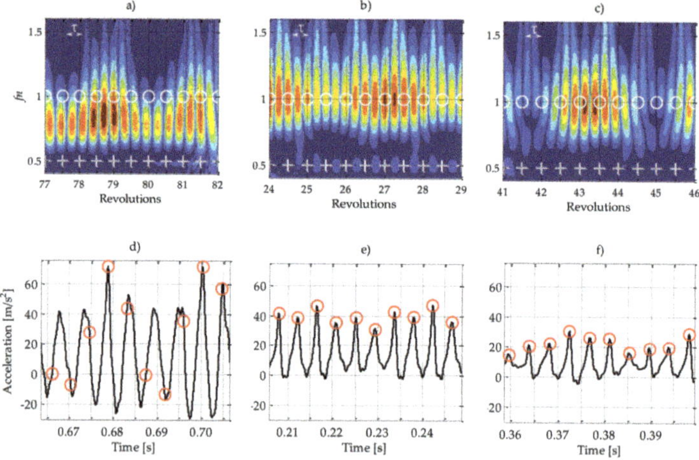

Figure 12. Analysis of the collected accelerometer signals (D, E and F) for different MR damper magnetic flux density values. Continuous wavelet transforms (**a,b,c**). Sampled acceleration signals (**d,e,f**).

5. Conclusions

Chatter vibration during the milling process of thin-floor components is a key factor that limits productivity. Placing an MR fluid in contact with the thin floor modifies the damping ratio without changing the natural frequency and modal stiffness. Experimental measurements indicate that a good degree of accuracy is attained from the proposed cutting force model, which is a key factor to predict stability lobes during the machining of thin floors using a bull-nose end mill.

- The proposed force model in which the tool edge is discretized in disks along the z-direction predicts the experimental cutting forces with high accuracy.
- The predicted EMHPM stability lobes of the cantilever beam closely follow experimental data.
- The use of an MR damper device located under the cantilever beam modifies modal damping depending on the magnitude of the magnetic flux density. For the higher magnetic flux density values, the damping ratio increases about 4 times, while the modal stiffness slightly varies.
- Under the effect of the MR damper device, the stability boundaries shift toward higher critical axial depth of cut values, which substantially enhances stable cutting conditions. In other words, experimental measurements indicate that when using the MR damper device, it is possible to increase the critical depth of cut from 0.5 to 1.5 mm in the range of spindle speed from 7000 to 10,000 rev/min, with an increase in the material removal rate and productivity by a factor of at least 3.

- The use of a semi-active MR damper device represents an alternative way to increase productivity while machining thin-floor components because of its robustness and adaptability to complex geometries. Since MR fluids can modify their yield and shear stresses as a function of the applied magnetic flux density, the damping ratio can be adjusted as needed in order to reach stable cutting conditions. Additionally, the semi-MR damper device is simple to operate versus its counterpart, the active MR damper device, which needs sensors and real-time data processing.

Author Contributions: Conceptualization, S.D.P.-A. and D.O.-T.; Methodology, S.D.P.-A.; Resources, O.M.-R., A.E.-Z. and L.N.L.d.L.; Supervision, O.M.-R., A.E.-Z. and L.N.L.d.L.; Validation, S.D.P.-A., D.O.-T. and G.U.; Writing—Original draft, S.D.P.-A. and D.O.-T.; Writing—Review and editing, O.M.-R., G.U. and A.E.-Z. All authors have read and agreed to the published version of the manuscript.

Funding: This research was funded by Tecnológico de Monterrey through the Research Group of Nanotechnology for Devices Design, and by the Consejo Nacional de Ciencia y Tecnología de México (Conacyt), Project Numbers 242269, 255837, 296176, and National Lab in Additive Manufacturing, 3D Digitizing and Computed Tomography (MADiT) LN299129.

Conflicts of Interest: The authors declare no conflict of interest. The founding sponsors had no role in the design of the study; in the collection, analyses, or interpretation of data; in the writing of the manuscript, and in the decision to publish the results.

References

1. Del Sol, I.; Rivero, A.; De Lacalle, L.N.L.; Gamez, A.; De Lacalle, L.N.L. Thin-Wall Machining of Light Alloys: A Review of Models and Industrial Approaches. *Materials* **2019**, *12*, 2012. [CrossRef] [PubMed]

2. Ahmed, G.S.; Reddy, P.R.; Seetharamaiah, N. Experimental Evaluation of Metal Cutting Coefficients under the Influence of Magneto-rheological Damping in End Milling Process. *Procedia Eng.* **2013**, *64*, 435–445. [CrossRef]

3. Junbai, L.; Kai, Z. Multi-point location theory, method, and application for flexible tooling system in aircraft manufacturing. *Int. J. Adv. Manuf. Technol.* **2010**, *54*, 729–736. [CrossRef]

4. Kalocsay, R.; Kolvenbach, C. Innoclamp GmbH–Hydraulic Clamping Systems. Available online: https://www.innoclamp.de/. (accessed on 28 June 2020).

5. Woody, S.C.; Smith, S.T. Damping of a thin-walled honeycomb structure using energy absorbing foam. *J. Sound Vib.* **2006**, *291*, 491–502. [CrossRef]

6. Park, G.; Bement, M.; Hartman, D.A.; Smith, R.E.; Farrar, C.R. The use of active materials for machining processes: A review. *Int. J. Mach. Tools Manuf.* **2007**, *47*, 2189–2206. [CrossRef]

7. Zhu, X.; Jing, X.; Cheng, L. Magnetorheological fluid dampers: A review on structure design and analysis. *J. Intell. Mater. Syst. Struct.* **2012**, *23*, 839–873. [CrossRef]

8. Symans, M.D.; Constantinou, M.C. Semi-active control systems for seismic protection of structures: A state-of-the-art review. *Eng. Struct.* **1999**, *21*, 469–487. [CrossRef]

9. Ji, H.; Huang, Y.; Nie, S.; Yin, F.; Dai, Z. Research on Semi-Active Vibration Control of Pipeline Based on Magneto-Rheological Damper. *Appl. Sci.* **2020**, *10*, 2541. [CrossRef]

10. Díaz-Tena, E.; Marcaide, L.L.D.L.; Gómez, F.C.; Bocanegra, D.C. Use of Magnetorheological Fluids for Vibration Reduction on the Milling of Thin Floor Parts. *Procedia Eng.* **2013**, *63*, 835–842. [CrossRef]

11. Fleischer, J.; Denkena, B.; Winfough, B.; Mori, M. Workpiece and Tool Handling in Metal Cutting Machines. *CIRP Ann.* **2006**, *55*, 817–839. [CrossRef]

12. Segalman, D.; Redmond, J. Chatter suppression through variable impedance and smart fluids. *SMART Struct. Mater.* **1996**, *53*, 1689–1699.

13. Wang, M.; Fei, R. Chatter suppression based on nonlinear vibration characteristic of electrorheological fluids. *Int. J. Mach. Tools Manuf.* **1999**, *39*, 1925–1934. [CrossRef]

14. Mei, D.; Kong, T.; Shih, A.J.; Chen, Z. Magnetorheological fluid-controlled boring bar for chatter suppression. *J. Mater. Process. Technol.* **2009**, *209*, 1861–1870. [CrossRef]

15. Çeşmeci, Ş.; Engin, T. Modeling and testing of a field-controllable magnetorheological fluid damper. *Int. J. Mech. Sci.* **2010**, *52*, 1036–1046. [CrossRef]

16. Som, A.; Kim, N.-H.; Son, H. Semiactive Magnetorheological Damper for High Aspect Ratio Boring Process. *IEEE/ASME Trans. Mechatron.* **2015**, *20*, 1–8. [CrossRef]

17. Kishore, R.; Choudhury, S.K.; Orra, K. On-line control of machine tool vibration in turning operation using electro-magneto rheological damper. *J. Manuf. Process.* **2018**, *31*, 187–198. [CrossRef]

18. Zhang, Y.; Wereley, N.M.; Hu, W.; Hong, M.; Zhang, W. Magnetic Circuit Analyses and Turning Chatter Suppression Based on a Squeeze-Mode Magnetorheological Damping Turning Tool. *Shock Vib.* **2015**, *2015*, 1–7. [CrossRef]

19. Ma, J.; Zhang, D.; Wu, B.; Luo, M.; Liu, Y. Stability improvement and vibration suppression of the thin-walled workpiece in milling process via magnetorheological fluid flexible fixture. *Int. J. Adv. Manuf. Technol.* **2016**, *88*, 1231–1242. [CrossRef]

20. Olvera, D.; Urbicain, G.; Elías-Zúñiga, A.; De Lacalle, L.N.L.; De Lacalle, L.N.L. Improving Stability Prediction in Peripheral Milling of Al7075T6. *Appl. Sci.* **2018**, *8*, 1316. [CrossRef]

21. Campa, F.J.; De Lacalle, L.L.; Celaya, A.; De Lacalle, L.N.L. Chatter avoidance in the milling of thin floors with bull-nose end mills: Model and stability diagrams. *Int. J. Mach. Tools Manuf.* **2011**, *51*, 43–53. [CrossRef]

22. Budak, E.; Altintaş, Y.; Armarego, E.J.A. Prediction of Milling Force Coefficients From Orthogonal Cutting Data. *J. Manuf. Sci. Eng.* **1996**, *118*, 216–224. [CrossRef]

23. Lee, P.; Altintaş, Y. Prediction of ball-end milling forces from orthogonal cutting data. *Int. J. Mach. Tools Manuf.* **1996**, *36*, 1059–1072. [CrossRef]

24. Yücesan, G.; Altintas, Y. Prediction of Ball End Milling Forces. *J. Eng. Ind.* **1996**, *118*, 95–103. [CrossRef]

25. Altintas, Y. Analytical Prediction of Three Dimensional Chatter Stability in Milling. *JSME Int. J. Ser. C* **2001**, *44*, 717–723. [CrossRef]

26. Olvera, D.; Elías-Zúñiga, A.; Martínez-Alfaro, H.; De Lacalle, L.L.; Rodríguez, C.A.; Campa, F.J.; De Lacalle, L.N.L. Determination of the stability lobes in milling operations based on homotopy and simulated annealing techniques. *Mechatronics* **2014**, *24*, 177–185. [CrossRef]

27. Urbikain, G.; Olvera, D.; De Lacalle, L.N.L.; Urbicain, G. Stability contour maps with barrel cutters considering the tool orientation. *Int. J. Adv. Manuf. Technol.* **2016**, *89*, 2491–2501. [CrossRef]

28. Smith, K.; Dvorak, D. Tool path strategies for high speed milling aluminum workpieces with thin webs. *Mechatronics* **1998**, *8*, 291–300. [CrossRef]

29. Seguy, S.; Campa, F.J.; De Lacalle, L.N.L.; Arnaud, L.; Dessein, G.; Aramendi, G. Toolpath dependent stability lobes for the milling of thin-walled parts. *Int. J. Mach. Mach. Mater.* **2008**, *4*, 377. [CrossRef]

30. Campa, F.J.; De Lacalle, L.N.L.; Lamikiz, A.; Sanchez, J.A.; Lamikiz, A. Selection of cutting conditions for a stable milling of flexible parts with bull-nose end mills. *J. Mater. Process. Technol.* **2007**, *191*, 279–282. [CrossRef]

31. Altıntaş, Y.; Lee, P. Mechanics and Dynamics of Ball End Milling. *J. Manuf. Sci. Eng.* **1998**, *120*, 684–692. [CrossRef]

32. Altintas, Y. Manufacturing Automation. 2011. *Appl. Mech. Rev.* **2001**, *54*, B84.

33. Compeán, F.; Olvera, D.; Campa, F.J.; De Lacalle, L.L.; Elías-Zúñiga, A.; Rodríguez, C.A.; De Lacalle, L.N.L. Characterization and stability analysis of a multivariable milling tool by the enhanced multistage homotopy perturbation method. *Int. J. Mach. Tools Manuf.* **2012**, *57*, 27–33. [CrossRef]

34. Olvera-Trejo, D.; Zúñiga, A.; De Lacalle, L.N.L.; Rodríguez, C.A. Approximate Solutions of Delay Differential Equations with Constant and Variable Coefficients by the Enhanced Multistage Homotopy Perturbation Method. *Abstr. Appl. Anal.* **2015**, *2015*, 1–12. [CrossRef]

35. Insperger, T.; Stepan, G. Updated semi-discretization method for periodic delay-differential equations with discrete delay. *Int. J. Numer. Methods Eng.* **2004**, *61*, 117–141. [CrossRef]

36. Mei, Y.; Mo, R.; Sun, H.; He, B.; Bu, K. Stability Analysis of Milling Process with Multiple Delays. *Appl. Sci.* **2020**, *10*, 3646. [CrossRef]

37. Insperger, T.; Stepan, G.; Bayly, P.V.; Mann, B. Multiple chatter frequencies in milling processes. *J. Sound Vib.* **2003**, *262*, 333–345. [CrossRef]

38. Lord Corp, MRF-122EG Magneto-Rheological Fluid. Available online: http://www.lordmrstore.com/lord-mr-products/mrf-122eg-magneto-rheological-fluid. (accessed on 29 June 2020).

Article

Double B-Spline Curve-Fitting and Synchronization-Integrated Feedrate Scheduling Method for Five-Axis Linear-Segment Toolpath

Xiangyu Gao, Shuyou Zhang, Lemiao Qiu *, Xiaojian Liu, Zili Wang and Yang Wang

State Key Laboratory of Fluid Power and Mechatronic Systems, Zhejiang University, Hangzhou 310027, China; gxy14@zju.edu.cn (X.G.); zsy@zju.edu.cn (S.Z.); liuxj@zju.edu.cn (X.L.); ziliwang@zju.edu.cn (Z.W.); onward@zju.edu.cn (Y.W.)

* Correspondence: qiulm@zju.edu.com; Tel.: +86-138-5800-2332

Received: 6 April 2020; Accepted: 28 April 2020; Published: 1 May 2020

Abstract: The discontinuities of a five-axis linear-segment toolpath result in fluctuation in the feedrate, acceleration and jerk commands that lead to machine tool vibration and poor surface finish. For path smoothing, with the global curve-fitting method it is difficult to control fitting error and the local corner-smoothing method has large curvature extreme. For path synchronization, the parameter synchronization method cannot ensure smooth rotary motion. Aiming at these problems, this paper proposes a double B-spline curve-fitting and synchronization-integrated feedrate scheduling method. Two C2-continuous and error-bounded B-spline curves are produced to fit tool-tip position and tool orientation, respectively. The fitting error is controlled by locally refining the curve segments that exceed the fitting tolerance. The tool-tip position trajectory is firstly planned to address axial kinematic constraints in the feedrate scheduling process. Then the feedrate is deformed for the tool orientation to guarantee smooth rotary motion as well as to share the same motion time with the tool-tip position segment by segment. The feasibility and effectiveness of the proposed method have been validated by simulations and experiments on the S-shape test piece. Simulations show that the proposed curve-fitting method can generate smooth toolpath and constrain fitting error. The proposed feedrate scheduling method can guarantee smooth rotary motion and keep axial motions under kinematic limits, compared with the method that does not consider axial kinematic constraints and the parameter synchronization method. Experimental results verify that the proposed curve-fitting method can generate smooth tool path under fitting tolerance, and the proposed feedrate scheduling method can produce smooth and restricted axial motions.

Keywords: five-axis linear-segment toolpath; path smoothing; B-spline curve-fitting; path synchronization; feedrate scheduling

---

1. Introduction

To machine free-form parts used widely in aerospace, die and mould industries, cutter of the machine tool is required to follow the programmed toolpath under computer numerical control (CNC) commands [1,2]. The curvilinear toolpath generated by computer-aided manufacturing (CAM) software is usually approximated by successive linear segments (linear-segment toolpath or G01 blocks) [3]. However, the transition corners between adjacent linear segments lead to tangential and curvature discontinuities, which cause fluctuation in the feedrate, acceleration, and jerk commands in the feedrate scheduling process that generates machine tool vibration and poor surface finish [4]. To avoid this behavior, the continuity of the linear-segment toolpath needs to be improved. Several approaches have been proposed in the existing literature for three-axis linear-segment toolpath smoothing, which can be categorized as local corner-smoothing method and global curve-fitting method. The local corner-smoothing method utilizes parametric curves to locally replace the sharp corners, such as B-spline [5–7], Bezier [8–10], and Pythagorean-hodograph (PH) curves [11,12]. Zhang et al. [7] propose G4-continuous toolpath via a quintic B-spline curve. Optimized by curvature variation energy (CVE), the chord error is seriously constrained and curvature extrema can be analytically calculated. Then the jerk-smooth feedrate scheduling scheme is developed based on the bidirectional scanning algorithm. Fan et al. [10] generated a G4 interpolative trajectory model using a symmetric nine-degree Bezier curve with confined chord error and analytical curvature extrema for trajectory smoothing, and employs jerk-smooth feedrate mode to perform time-optimal feedrate scheduling. Shi et al. [12] adopt a curvature-continuous PH curve as a transition to blend sharp corners. The blending algorithm can guarantee the approximation error exactly. The control points, the arc length and the curvature of the transition curve also have analytical expressions. The high-order geometric continuity leads to high-smoothness toolpath and feedrate profile. But the rising complexity may prevent these methods from promoting to five-axis processing. On the other hand, the global curve-fitting method utilizes a parametric curve to smooth the linear-segment toolpath globally by interpolating or approximating G01 data points. However, this method suffers from below-standard fitting error. Bi et al. [13] propose a general, fast and robust B-spline fitting scheme with PIA method to generate G2 tool path under confined chord error for high speed interpolation of micro-line tool path. He et al. [14] propose a progressive and iterative approximation for least squares incorporating energy term (ELSPIA) to deal with chord error constraint. Lin et al. [15] derive the explicit expression for the Hausdorff distance between a line segment and a curve segment, then a time-parameterized curve-fitting algorithm is presented to combine path smoothing and trajectory planning.

Five-axis CNC machine tools provide better accessibility with the help of additional rotary axes [16,17]. The smoothing of five-axis linear-segment toolpath brings about extra difficulties compared to three-axis. First, the tool orientation must be smoothed to ensure smooth rotary motion in addition to the tool-tip position smoothing. Second, given that the tool-tip position and tool orientation are smoothed independently, the two smoothed trajectories must be synchronized [18]. The local corner-smoothing method and the global curve-fitting method mentioned above are extended and used to smooth tool-tip position and tool orientation respectively. The local corner-smoothing method applied to five-axis linear-segment toolpath can guarantee high-order continuity while respect error tolerance and synchronization of the tool-tip position and tool orientation is achieved by adjusting the joint points of the linear and curvilinear segments [19–21]. The global curve-fitting method applied to five-axis linear-segment toolpath can ensure toolpath smoothness, but the fitting error is hard to control and evaluate [22,23]. Moreover, the global curve-fitting method usually synchronizes two trajectories by sharing the same curve parameter (parameter synchronization). The parameter synchronization method [22,23] plans the tool-tip trajectory according to translational motion constraints and then makes the tool orientation follow the tool-tip in the curve parameter space. However, if the tool orientation is merely parameter synchronized with tool-tip position, the rotary motion may be discontinuous at certain locations.

After path smoothing and synchronization, the geometric information of toolpath is transferred to the feedrate, acceleration (the change rate of feedrate) and jerk (the change rate of acceleration) trajectories in CNC system [24]. The toolpath with high-order continuity can raise the feedrate at the sharp corner, decline the fluctuation of feedrate and acceleration, thus improve the machining quality. Many feedrate scheduling methods have been proposed in the literature, including the jerk-limited method [25–27], the optimization method [28], the linear programming method [29–31], and the time-synchronization method [32]. Du et al. [25] present a complete S-shape feedrate scheduling approach (CSFA) with limited jerk, acceleration and feedrate for the three-axis parametric toolpath. Beudaert et al. [28] propose a Velocity Profile Optimization (VPOp) algorithm which has been implemented for various toolpath format. Fan et al. [29] reduce the velocity planning problem to an equivalent linear programming problem with polynomial computational complexity O(N3.5) to find the optimal solution. Huang et al. [32] advise a real-time feed scheduling method for five-axis machining by simultaneously planning linear and angular trajectories (SLATP) considering axial kinematic constraints.

For path smoothing, the global curve-fitting method has difficulty in controlling fitting error and the local corner-smoothing method has large curvature extreme which decreases nominal velocity in feedrate scheduling. For path synchronization, the parameter synchronization method cannot guarantee smooth rotary motion. With the aim of resolving these problems, a double B-spline curve-fitting and synchronization-integrated feedrate scheduling method for five-axis linear-segment toolpath is presented in this paper. Two cubic B-spline curves are used for fitting the tool-tip position and tool orientation respectively. The data points are initially fitted by a B-spline curve and then the curve segments exceeding the fitting tolerance are locally refined to get the final curve. After curve-fitting, synchronization and feedrate scheduling is required to guarantee smooth motion and machining efficiency. Feedrate scheduling of the tool-tip position is conducted first to address the axial kinematic constraints. Then the feedrate of the tool orientation is deformed to share the same motion time with the tool-tip position so that the tool-tip position and tool orientation arrive at specific locations simultaneously.

The rest of the paper is organized as follows. The fitted curves for tool-tip position and tool orientation under fitting tolerance are generated in Section 2. The synchronization and feedrate scheduling of the tool-tip position and tool orientation are implemented in Section 3. Simulations and experiments are carried out in Section 4 to demonstrate the feasibility and effectiveness of the proposed method. Conclusions are drawn in Section 5.

2. Double B-Spline Curve-Fitting for Tool-Tip Position and Tool Orientation under Fitting Tolerance

In this section, a double B-spline curve-fitting method is proposed to smooth tool-tip position and tool orientation in different coordinate frames respectively. The overview of the proposed curve-fitting method is shown in Figure 1. Five-axis linear-segment toolpath is expressed as discrete cutter location (CL) data. CL data in the workpiece coordinate system (WCS) are defined as $[\mathbf{p}_i, \mathbf{o}_i]^T$, where the Cartesian coordinate $\mathbf{p}_i = [p_{ix}, p_{iy}, p_{iz}]^T$ represents the tool-tip position, and the spherical coordinate $\mathbf{o}_i = [o_{iI}, o_{iJ}, o_{iK}]^T$ represents the tool orientation. The tool orientation spherical coordinate can be mapped into $[\alpha, \beta]^T$ plane, where α and β are two rotary angles in the machine coordinate system (MCS).

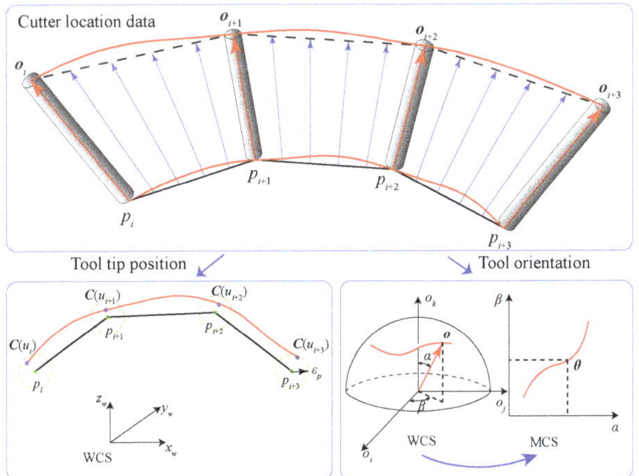

Figure 1. Overview of double B-spline curve-fitting.

2.1. Tool-Tip Position Curve-Fitting

To fit successive linear segments, a cubic B-spline curve is used as the initial curve to guarantee C2-continuity (C2-continuity is necessary for continuous acceleration in feedrate scheduling). Besides geometric continuity, the fitting error and curvature are taken into consideration as criteria of a fitted curve. The fitting error affects the dimensional accuracy of the machined workpiece, hence must be regulated under fitting tolerance. The curvature, which affects the motion kinematics (the nominal velocity in feedrate scheduling), should be as small as possible but not under strict limitation.

The fitting error of tool-tip position is satisfied if the Hausdorff distance between the fitted curve and the tool-tip position polyline is not greater than the fitting tolerance ε_p in the WCS. For a curve segment $\mathbf{C}(s_{k-1})\mathbf{C}(s_k)$ and a line segment $\overline{\mathbf{p}_{k-1}\mathbf{p}_k}$, the Hausdorff distance is [15]:

$$
\begin{aligned}
\mathrm{Dis}_H\left(\overline{\mathbf{p}_{k-1}\mathbf{p}_k}, \mathbf{C}(s_{k-1})\mathbf{C}(s_k)\right) = \\
\max\Big\{ \max_{\mathbf{p}\in\overline{\mathbf{p}_{k-1}\mathbf{p}_k}}\{\mathrm{Dis}(\mathbf{p}, \mathbf{C}(s_{k-1})\mathbf{C}(s_k))\}, \max_{\mathbf{C}\in\mathbf{C}(s_{k-1})\mathbf{C}(s_k)}\{\mathrm{Dis}(\mathbf{C}, \overline{\mathbf{p}_{k-1}\mathbf{p}_k})\}\Big\},
\end{aligned}
\tag{1}
$$

where $\mathrm{Dis}(\mathbf{a}, \mathbf{b})$ is the Euclidean distance between $\mathbf{a}$ and $\mathbf{b}$.

Since the explicit calculation of the Hausdorff distance by its definition between the fitted curve and the polyline is very computation-complex, the fitting error is classified into two categories. The first is called point error that indicates the approximation error between the fitting curve and the data points. The second is called chord error that represents the approximation error between the fitting curve and the linear segments.

To fit the tool-tip position data, we start with a cubic B-spline curve to satisfy point error. A cubic B-spline curve defined as linear combination of control points $\mathbf{P}_i$ and cubic B-spline basis functions $N_{i,d}(u)$ is given by [33]:

$$
\mathbf{C}(u) = \sum_{i=0}^{m} N_{i,d}(u)\mathbf{P}_i, \quad 0 \le u \le 1,
\tag{2}
$$

where d stands for degree of the curve and in this case $d = 3$. The B-spline basis functions defined on a non-descending knot vector $\mathbf{U} = \{0, 0, 0, 0, u_4, \ldots, u_n, 1, 1, 1, 1\}$ are calculated recursively as follows [33]:

$$
\begin{cases}
N_{i,0}(u) = \begin{cases} 1, & u_i \le u < u_{i+1} \\ 0, & otherwise \end{cases} \\
N_{i,d}(u) = \frac{u-u_i}{u_{i+d}-u_i} N_{i,d-1}(u) + \frac{u_{i+d+1}-u}{u_{i+d+1}-u_{i+1}} N_{i+1,d-1}(u).
\end{cases}
\tag{3}
$$

The cubic B-spline curve that passes through a set of tool-tip position data points $\mathbf{p} = \{\mathbf{p}_i \in R^3 | i = 0, 1, \ldots, m\}$ is determined once the knot vector $\mathbf{U}$ and control points $\mathbf{P}_i$ are decided. The curve parameter value $\bar{u}_i$ corresponding to data point $\mathbf{p}_i$ and the knot vector $\mathbf{U}$ are responsible for the shape and parameterization of the B-spline curve. $\bar{u}_i$ is assigned using the cumulative chord length parameterization [33]:

$$
\begin{cases}
\bar{u}_0 = 0, \ \bar{u}_m = 1; \\
\bar{u}_k = \bar{u}_{k-1} + \frac{\|\mathbf{p}_k - \mathbf{p}_{k-1}\|}{L}, & k = 1, \ldots, m-1,
\end{cases}
\tag{4}
$$

where $L = \sum_{k=1}^{m} \|\mathbf{p}_k - \mathbf{p}_{k-1}\|$. The knot vector $\mathbf{U} = [u_0, u_1, \ldots, u_{n+4}]$ is built by averaging parameter values from Equation (3) as follows [33]:

$$
\begin{cases}
u_0 = \cdots = u_3 = 0, \ u_{n+1} = \cdots u_{n+4} = 1; \\
u_{j+3} = \frac{1}{3} \sum_{i=j}^{j+2} \bar{u}_i, & j = 1, 2, \ldots, n-3.
\end{cases}
\tag{5}
$$

Since the knot vector **U** has been decided and each tool-tip position point has been assigned a parameter value, system of linear equations is established:

$$\mathbf{p}_k = \mathbf{C}(\bar{u}_k) = \sum_{i=0}^{m} N_{i,3}(\bar{u}_k)\mathbf{P}_i, \quad k = 0,1,\ldots,m. \tag{6}$$

The control points are acquired by solving system of linear equations shown in Equation (6) using the inverse B-spline basis function matrix. Therefore, the initial curve interpolating all the data points defined by Equation (1) is resolved, and the point error is satisfied. However, the initial curve might exceed chord error at certain curve segments.

Assuming that the distance between the initial curve in the parameter interval $[\bar{u}_{k-1}, \bar{u}_k]$ and the corresponding line segment $\overline{\mathbf{p}_{k-1}\mathbf{p}_k}$ is greater than ε_p, the local refinement process is expected to be limited in the parameter interval so that the remaining portion of the curve is not affected.

In order to increase the flexibility of the local curve segment that needs to be refined, control points are added using knots insertion. Let u_j be the knot inserted to $[\bar{u}_{k-1}, \bar{u}_k]$, the new control points can be calculated as:

$$\begin{cases} \mathbf{c}_i = \mathbf{P}_i, & i = 0,\ldots,k-d-1 \\ \mathbf{c}_i = (1-\alpha_i)\mathbf{P}_{i-1} + \alpha_i\mathbf{P}_i, & i = k-d,\ldots,k-1 \\ \mathbf{c}_{i+1} = \mathbf{P}_i, & i = k,\ldots,m, \end{cases} \tag{7}$$

where $\alpha_i = (u_j - u_{k-1})/(u_{k+d-1} - u_{k-1})$.

Denote $\bar{u}_{k-1}$ and $\bar{u}_k$ as u_s and u_e after knot insertion. Denote the distance between the control point $\mathbf{P}_i(i = s-1,\ldots,e-p)$ and $\overline{\mathbf{p}_{k-1}\mathbf{p}_k}$ as D_i. D_{s-1} and D_{e-p} which must be not greater than ε_p due to strong convex hull property of B-spline curves [33], otherwise it is necessary to insert more knots to $[\bar{u}_{k-1}, \bar{u}_k]$. To keep the portion outside $[\bar{u}_{k-1}, \bar{u}_k]$ unchanged, we can only adjust the control points $\mathbf{P}_j(j = s,\ldots,e-p-1)$. If $D_j > \varepsilon_p$, replace $\mathbf{P}_j$ with $\mathbf{P}_j' = \mathbf{P}_j + \alpha_j\mathbf{W}$ where $\mathbf{W}$ is the unit vector that points from $\mathbf{P}_j$ to $\mathbf{P}_j'$ and α_j is the distance moved along $\mathbf{W}$. Assign $\mathbf{W}$ to the unit vector of the direction that is perpendicular to $\overline{\mathbf{p}_{k-1}\mathbf{p}_k}$ and α_j to the value of $D_j - \varepsilon_p$. After all the local curve segments are refined, the final curve satisfies the fitting tolerance while maintaining C2-continuity. Flow of the proposed curve-fitting scheme is shown in Figure 2.

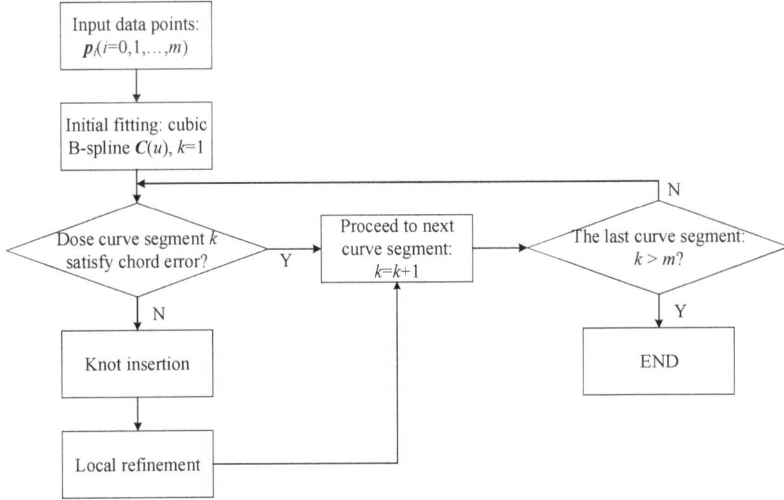

Figure 2. Flowchart of the proposed curve-fitting scheme.

We present a planar example with five points and fitting tolerance of 0.05 mm as a numerical illustration of the proposed curve-fitting scheme. The result of the B-spline curve fitted to five data points is shown in Figure 3. The initial curve that satisfies point error exceeds chord error bound at segment 1 and segment 2, which are zoomed in to get a clearer view. Then the two segments is locally refined to obtain the final error-constrained curve. The fitting error and curvature results are demonstrated in Figure 4. As can be seen, the fitting error of the traditional global curve-fitting method exceeds the fitting tolerance. The proposed curve-fitting method, as an improvement of the global curve-fitting method, can get the fitting error strictly under the fitting tolerance. Compare to the local corner-smoothing method, the proposed method has smaller curvature extrema under the same geometric continuity. Notice that the local corner-smoothing method has lower fitting error in most areas because only the corner parts are replaced. The proposed curve-fitting method is more adaptive to micro linear segments whose adjacent corners are very close.

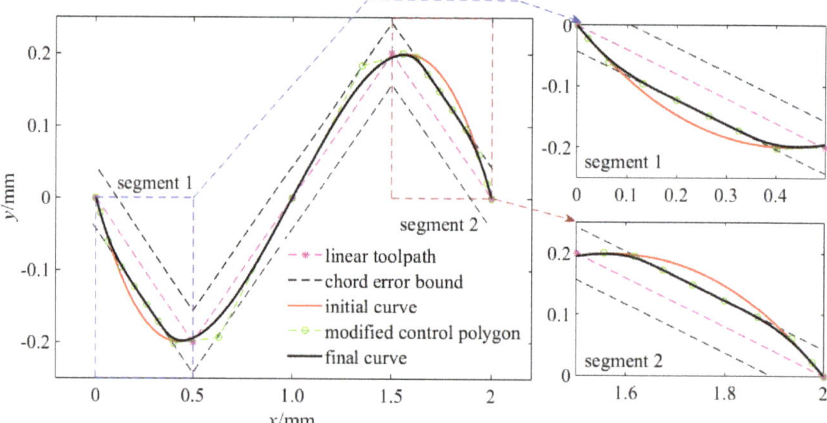

Figure 3. Fitted curve of planar five-point example.

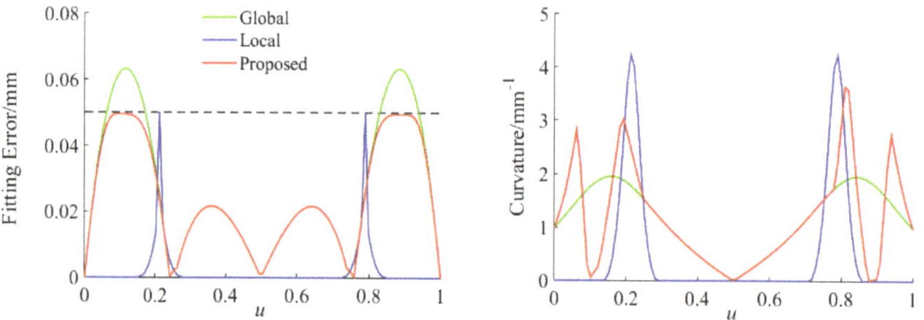

Figure 4. Fitting error and curvature of planar five-point example.

2.2. Tool Orientation Curve-Fitting

The tool orientation data $\mathbf{o} = \{\mathbf{o}_i \in R^3 | i = 0, 1, \ldots, m\}$ are expressed by the spherical coordinates $\mathbf{o}_i = [o_{iI}, o_{iJ}, o_{iK}]$ in the WCS. Unlike curve-fitting of the tool-tip position, the tool orientation is fitted to the Cartesian coordinates that map from the given spherical coordinates to ensure that the magnitude of the orientation vector remains unity [22]. The mapping from the spherical coordinates to the Cartesian coordinates is obtained as follows:

$$\begin{cases} \alpha_i = \cos^{-1}(o_{iK}) \\ \gamma_i = \tan^{-1}(o_{iI}/o_{iJ}), \end{cases} \tag{8}$$

where α_i and γ_i are actually the rotary drive commands θ_{Ai} and θ_{Ci} in the MCS respectively. And the inverse mapping is in the following form:

$$\begin{cases} o_{iI} = \sin\theta_{Ai}\sin\theta_{Ci} \\ o_{iJ} = \sin\theta_{Ai}\cos\theta_{Ci} \\ o_{iK} = \cos\theta_{Ai}. \end{cases} \tag{9}$$

The fitting tolerance of the tool orientation is given as ε_o in the WCS. Since the tool orientation is smoothed in the MCS, it is necessary to convert ε_o to ε_θ which denotes the fitting tolerance of the tool orientation in the MCS.

As shown in Figure 5, we assume O' is the orientation position after O rotates ε in the WCS and O_t is the transitional orientation position that rotates $\Delta\alpha$ away from the z_w-axis to reach O'. The lengths of edge $\overline{OO_t}$, $\overline{O_tO'}$, and $\overline{OO'}$ are obtained as:

$$\begin{cases} a = 2\sin\alpha\sin(0.5\Delta\beta) \\ b = 2\sin(0.5\Delta\alpha) \\ c = 2\sin(0.5\varepsilon). \end{cases} \tag{10}$$

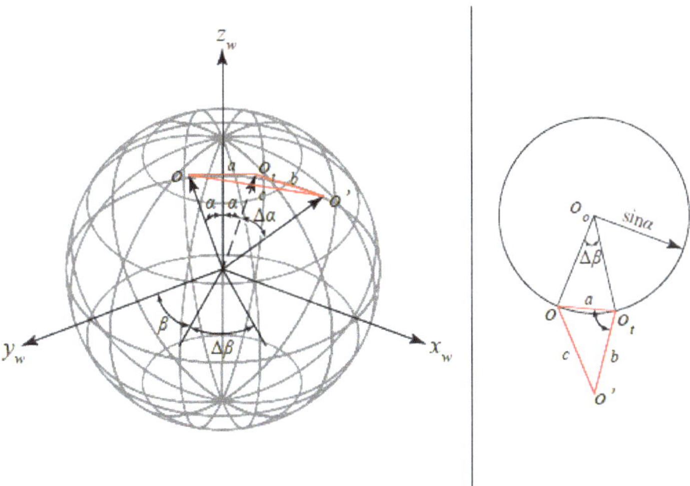

Figure 5. Fitting error and curvature of planar five-point example.

Since $\Delta\alpha$, $\Delta\beta$, and ε are extremely small, O_o, O_t, and O' are approximately collinear. Thus we have $\angle O_oO_tO \approx \angle OO_tO' \approx 90°$ and Equation (10) can be simplified as:

$$\begin{cases} a = (\sin\alpha)\Delta\beta \\ b = \Delta\alpha \\ c = \varepsilon. \end{cases} \tag{11}$$

Because $\angle OO_tO' \approx 90°$, we have $a^2 + b^2 = c^2$, that is $(\sin \alpha)^2 \Delta \beta^2 + \Delta \alpha^2 = \varepsilon^2$ which lead to $\varepsilon^2 \leq \Delta \alpha^2 + \Delta \beta^2 = \varepsilon_\theta^2$. Since $\varepsilon \leq \varepsilon_0$, the fitting error of the tool orientation is satisfied if $\varepsilon_\theta \leq \varepsilon_0$. Without loss of generality, let $\varepsilon_\theta = \varepsilon_0$.

The B-spline curve in the $\theta_A \theta_C$ Cartesian plane is obtained with the input of the mapped tool orientation data $\boldsymbol{\theta} = \{(\theta_{Ai}, \theta_{Ci}) | i = 0, 1, \ldots, m\}$ and fitting tolerance ε_θ according to the procedure in Section 2.1 as:

$$\mathbf{R}(v) = \sum_{i=0}^{n} N_{i,d}(v)\boldsymbol{\theta}_i. \tag{12}$$

The tool orientation spline $\mathbf{o}(v) = [o_I(v), o_J(v), o_K(v)]$ in the WCS is obtained using inverse mapping from Equation (9). It is noteworthy that the number of control points of the fitted curve depends on the initial curve and the fitting tolerance. Therefore the number of control points of the tool orientation is not necessarily the same as the tool-tip position.

3. Synchronization-Integrated Feedrate Scheduling

Since the tool-tip position and tool orientation are fitted independently in different coordinates, the synchronization and feedrate scheduling are necessary for curve interpolation. In this section, synchronization-integrated feedrate scheduling method is presented to ensure that the tool-tip position motion and tool orientation motion are synchronized and smooth. Feedrate scheduling of the tool-tip position is conducted firstly to address the axial kinematic constraints. Then the tool orientation feedrate is obtained by feedrate deformation method to guarantee smooth rotary motion as well as to share the same motion time with the tool-tip position segment by segment. Thus, the synchronization and feedrate scheduling is implemented in one procedure.

3.1. Feedrate Scheduling for Tool-Tip Position

The goal of feedrate scheduling is to find a smooth interpolation under machining constraints. Usually, the interpolation chord error, the axial kinematics and the tangential kinematics are important constraints for high speed and high accuracy machining. Under the circumstances, the interpolation error, the axial velocities and accelerations are considered to determine the allowable feedrate along the tool-tip position curve $\mathbf{C}(u)$. Then the feedrate is scheduled using the S-shape jerk-limited method taking tangential acceleration and jerk into account.

The interpolation chord error along $\mathbf{C}(u)$ at the parameter value u can be approximated as [25]:

$$\delta(u) = \rho(u) - \sqrt{\rho(u) - (f(u)T/2)^2}, \tag{13}$$

where T is the interpolation period, $\rho(u)$ is the curvature radius, and $f(u)$ is the feedrate at the parameter value u. Hence the interpolation chord error limited feedrate can be obtained as:

$$f(u) \leq 2\sqrt{2\rho(u)\delta_{max} - \delta_{max}^2} \Big/ T, \tag{14}$$

where δ_{max} is the defined limitation of interpolation error. Further, the feedrate in the curve parameter domain can be expressed as:

$$f(u) = \left\| \frac{d\mathbf{C}(u)}{dt} \right\| = \left\| \frac{d\mathbf{C}(u)}{du} \right\| \frac{du}{dt}. \tag{15}$$

Denote $p = du/dt$ as a representation of transition between time domain t and curve parameter domain u, and Equation (14) can be converted to:

$$p(u) \leq 2\sqrt{\left(2\rho(u)\delta_{max} - \delta_{max}^2\right)} \Big/ \|\mathbf{C}_u(u)\| T. \tag{16}$$

The velocity of each axis can be calculated as:

$$v_M(u) = \frac{dM(u)}{dt} = \frac{dM(u)}{du}\frac{du}{dt} = M_u(u)p, \tag{17}$$

where $M(u)$ represents the drive displacement. If the maximum velocity of each axis is set as V_{max}^M, the axial velocity constraint is written as:

$$p(u) \leq \frac{V_{max}^M}{M_u(u)}. \tag{18}$$

The acceleration of each axis can be calculated as:

$$a_M(u) = \frac{dv_M}{dt} = \frac{dv_M}{dM}\frac{dM}{dt} = v_M\frac{dv_M}{dM} = \frac{1}{2}\frac{d(v_M^2)}{dM} = \frac{1}{2}\frac{d(pM_u)}{dM}. \tag{19}$$

If the maximum velocity of each axis is set as A_{max}^M, the axial velocity constraints can be written as:

$$\frac{d(pM_u)}{dM} \leq 2A_{max}^M. \tag{20}$$

The parameter interval of the tool-tip curve $C(u)$ is divided into N equal sub-intervals at $u_i = i/N(i = 0,1,\ldots,N)$. Using the monotonic rational quadratic spline proposed by Fleisig and Spence [34], the discrete parameter value v_i of the tool orientation corresponding to u_i is computed. When N is large enough, the length of the parameter sub-interval is so small that M_u in Equations (17) and (19) can be approximated using the finite differential method. We can address the interpolation chord error, the axial velocities and the axial acceleration constraints as an optimization problem with the objective of $\max \sum_{i=0}^{N} p_i$ and constraints from Equations (16), (18) and (20). By solving the optimization problem, the allowable discrete feedrate values $f_i = \|C_u(u_i)\| p_i$ is acquired. Then the feedrate of the tool-tip position is generated utilizing the curve splitting and bi-directional scanning techniques.

3.2. Feedrate Deformation for Tool Orientation to Achieve Synchronization

The tool-tip position curve at the curve segment joint parameters $\bar{u}_i, i = 0,\ldots,m$ and the tool orientation curve at the parameter $\bar{v}_i, i = 0,\ldots,m$ are synchronized spontaneously. Since the feedrate scheduling of the tool-tip position is completed, the motion time t_k along the tool-tip curve $C(u)$ in the parameter interval $[\bar{u}_{k-1}, \bar{u}_k]$ is fixed. Thus the motion time along the tool orientation curve in the parameter interval $[\bar{v}_{k-1}, \bar{v}_k]$ must be equal to t_k to achieve synchronization. The tangential acceleration and jerk are set to zero at the joint parameters so that the tool-tip position and tool orientation motion are smooth.

Based on these principles, a feedrate deformation method is presented to generate the feedrate of the tool orientation. We take the tool-tip feedrate profile with seven phases in as an illustration. The durations of the seven phases are computed by the start and end feedrate v_s and v_e, the maximum feedrate v_{max}, the tangential acceleration a_{max} and jerk J_{max} and the arc-length L_p:

$$\begin{cases} t_1 = a_{max}/J_{max} \\ t_2 = (v_{max} - v_s)/a_{max} - t_1 \\ t_3 = t_1 \\ t_4 = [L_p - (v_{max} + v_s)T_a - (v_{max} + v_e)T_d]/v_{max} \\ t_5 = a_{max}/J_{max} \\ t_6 = (v_{max} - v_e)/a_{max} - t_5 \\ t_7 = t_5, \end{cases} \tag{21}$$

where $T_a = 2t_1 + t_2$ represents the acceleration time, $T_d = 2t_5 + t_6$ represents the deceleration time. Denote the constant feedrate duration t_4 as T_c. The feed displacement is calculated as:

$$L_p = \frac{v_s + v_{max}}{2} T_a + v_{max} T_c + \frac{v_{max} + v_e}{2} T_d. \tag{22}$$

The goal of feedrate deformation is to generate a tool orientation feedrate profile segment by segment with the same motion time as tool-tip position and nominal orientation displacement while keep tangential kinematics under constraint. Given the tool orientation start feedrate w_s, the tool orientation segmented arc-length L_o, the feedrate deformation begins with computing the new constant feedrate derived form Equation (22):

$$w_m = (2L_o - w_s T_a)/(T_a + 2T_c + (1 + k)T_d), \tag{23}$$

where $k = w_e/w_m$ is user-defined and T_a, T_c and T_d are the same as tool-tip position. By adjusting k, different constant feedrates can be obtained to meet the tangential constraints. The updated tangential acceleration and jerk is calculated as:

$$\begin{cases} a_s = (w_m - w_s)/(t_1 + t_2) \\ J_s = a_s/t_1 \\ a_s = (w_m - w_e)/(t_5 + t_6) \\ J_e = a_e/t_5. \end{cases} \tag{24}$$

If J_s or a_s still exceed the tangential constraint, we can consider stretching the acceleration time to $T'_a = T_a + \Delta t$ and then compute a new constant feedrate so that J_s and a_s are reduced. The deceleration process can be handled in the same way. Figure 6 is an illustration of the feedrate deformation method. The orientation feedrate profile is obtained from deformation of the position feedrate profile so that they share the same motion time but get through respective displacements to reach a specific synchronized cutter location.

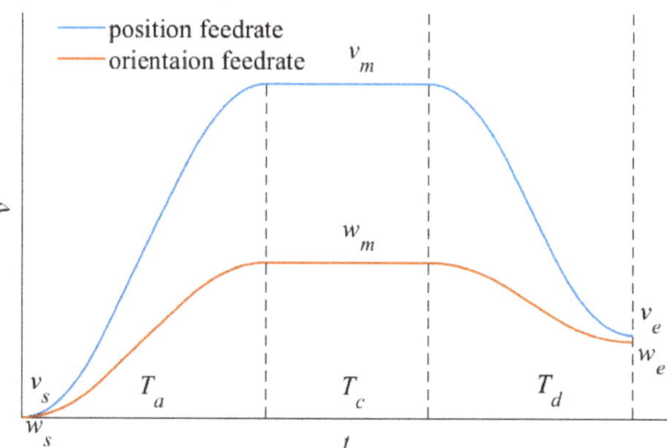

Figure 6. Illustration of feedrate deformation method.

With the new constant feedrate, tangential acceleration and jerk, the tool orientation feed time is adjusted to match the tool-tip position feed time while the segmented arc-length remains the same. As a consequence, the tool-tip position and tool orientation arrive at specific points simultaneously where the two curves are synchronized.

4. Simulation and Experiment Validations

The feasibility and effectiveness of the proposed method is validated by simulations on a computer with 4.0 GHz CPU and 8GB RAM and experiments on a five-axis table-tilting machine tool (model of the machine tool) with two rotary axes of A and C whose configuration is illustrated in Figure 7. The inverse kinematic transformation of the experimental machine tool is:

$$\begin{cases} A = \arccos(o_k) \\ C = \arctan(o_i/o_j) \\ X = -x\cos C - y\sin C \\ Y = x\cos A \sin C - y\cos A \sin C - z\sin A - L_{AC,z}\sin A \\ Z = x\sin A \sin C - y\sin A \cos C + z\cos A + L_{AC,z}\cos A, \end{cases} \tag{25}$$

where $L_{AC,z} = 40$ mm is the offset distance between A-axis and C-axis in the direction of z-axis. The machine is controlled by a TwinCAT-based CNC system with the sampling period of 1 ms. In order to formulate the feedrate profile to S-shape, we add the maximum allowable feedrate for tool-tip position and tool orientation respectively to the predefined constraints which are listed in Table 1.

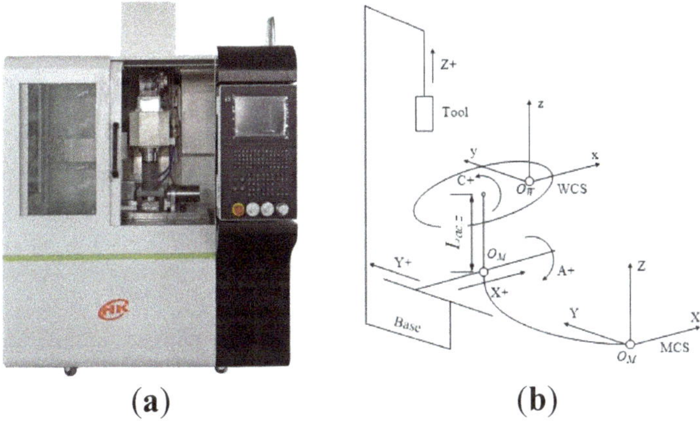

(a) **(b)**

Figure 7. Configuration of experimental five-axis computer numerical control (CNC) machine tool— (a) Real product. (b) Layout.

Table 1. Predefined constraints in the experiment.

Constraints	Values
Interpolation chord error	1 μm
X/Y/Z-axis velocity	100 mm/s
X/Y/Z-axis acceleration	1000 mm/s^2
A/C-axis velocity	0.5 rad/s
A/C-axis acceleration	5 rad/s^2
Tool-tip position tangential acceleration	200 mm/s^2
Tool-tip position tangential jerk	2000 mm/s^3
Tool orientation tangential acceleration	5 rad/s^2
Tool orientation tangential jerk	50 rad/s^3
Tool-tip position maximum feedrate	50 mm/s
Tool orientation maximum feedrate	0.5 rad/s

The S-shape test piece is used to analyze the performance of the proposed method. The S-shape test piece is defined by two ruled surfaces and is usually used as a representation of the thin-wall workpiece in the aerospace industry. It requires an allowable range of final error from -0.05 mm to $+0.05$ mm [35]. We import the CAD model of the test piece shown in Figure 8a to UG NX 10 to plan the toolpath of the ruled surfaces using five-axis flank milling. The programmed toolpath shown in Figure 8b is approximated by a series of discrete G01 data points. The five-axis linear-segment toolpath of one of the four corners which consists of 12 G01 data points shown in Figure 8c is used to demonstrate the effectiveness for the proposed curve-fitting and feedrate scheduling method. The experimental 12 G01 data points is available in Appendix A Table A1. The tool-tip position fitting error tolerance ε_p is 0.05 mm and the tool orientation fitting error tolerance ε_o is $0.05°$.

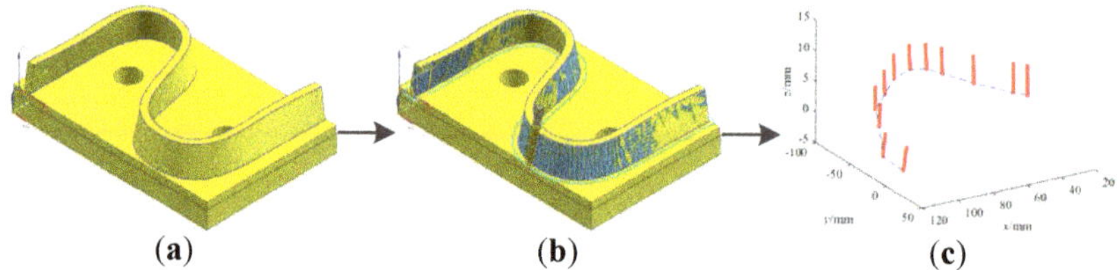

Figure 8. Linear-segment toolpath generation for S-shape test piece—(**a**) computer-aided design (CAD) model. (**b**) Toolpath generated by computer-aided manufacturing (CAM). (**c**) Experimental G01 CL data.

The results of curve-fitting for the 12-point linear-segment toolpath are shown in Figure 9 and the corresponding fitting error calculated numerically in the curve parameter space is shown in Figure 10. From the data, it is evident that the fitting errors of the tool-tip position and of the tool orientation are both under fitting tolerance.

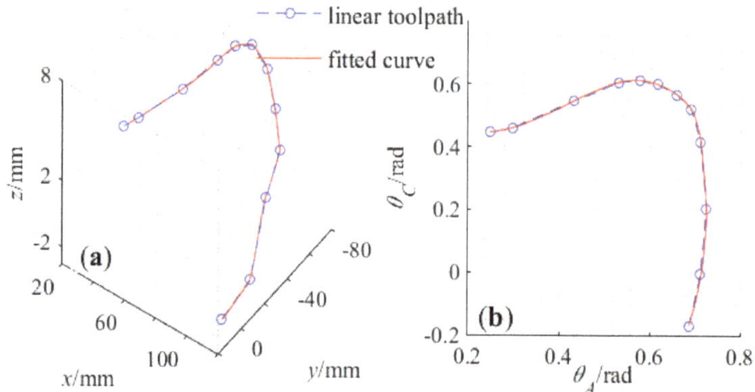

Figure 9. Results of fitted curve for the experimental linear-segment toolpath—(**a**) Tool-tip position. (**b**) Tool orientation.

We assign $N = 500$ to compute the discrete locations of the tool-tip position and tool orientation. Then the optimization problem proposed in Section 3.1 is established and solved using the *linprog* solver in Matlab to get the allowable feedrate for the tool-tip position. The feedrate profile of the tool-tip position is obtained using the curve splitting and bi-directional scanning techniques and is shown in Figure 11. From the data we can tell that the scheduled feedrate profile is smooth and under feedrate restrictions.

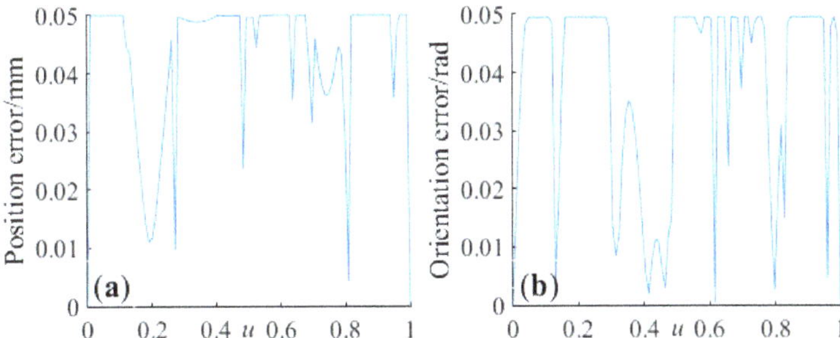

Figure 10. Fitting error of tool-tip position and tool orientation—(**a**) Tool-tip position. (**b**) Tool orientation.

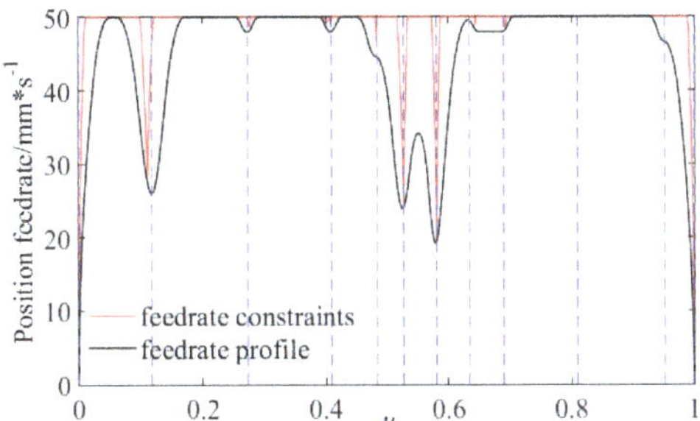

Figure 11. Feedrate profile of the tool-tip position with respect to curve parameter.

The feedrate profiles of the original linear-segment toolpath and the smooth toolpath are compared to show that the total motion duration is reduced by 44.2% from 7.169 s to 3.999 s in Figure 12. It can also be seen that the tool orientation feedrate is generated sharing the same segmented motion time with the tool-tip position.

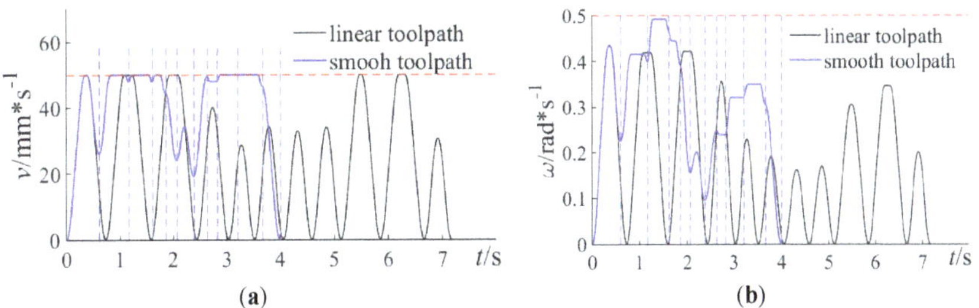

Figure 12. Feedrate profile with respect to time—(**a**) Tool-tip position. (**b**) Tool orientation.

The interpolation is conducted to generate reference positions in the WCS, as shown in Figure 13. From the data, it is evident that the feed process is advanced at high rate except three obtrusive deceleration locations.

The tangential kinematic profiles of the tool-tip position and tool orientation are illustrated in Figure 14. It can be spotted that all the tangential kinematic profiles meet with the tangential constraints. In Figure 14c,d, the results of the adjusting techniques mentioned in Section 3.2 is displayed in the feedrate deformation process.

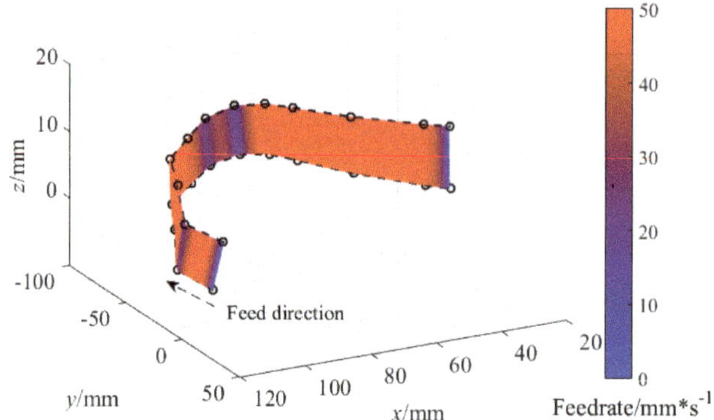

Figure 13. Fitting error and curvature of planar five-point example.

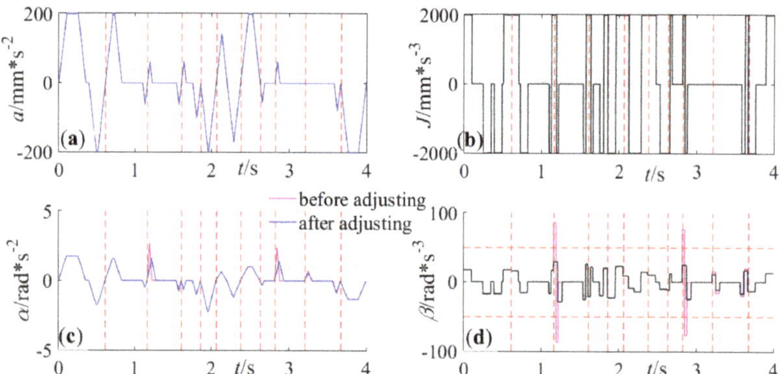

Figure 14. Tangential kinematic profiles—(**a**) Tool-tip position acceleration. (**b**) Tool-tip position jerk. (**c**) tool orientation acceleration. (**d**) tool orientation jerk.

The traditional parameter synchronization method obtains the tool orientation feedrate after scheduling the tool-tip feedrate by simply sharing the curve parameter. It may cause wild fluctuation in the tool orientation kinematics, as shown in Figure 15.

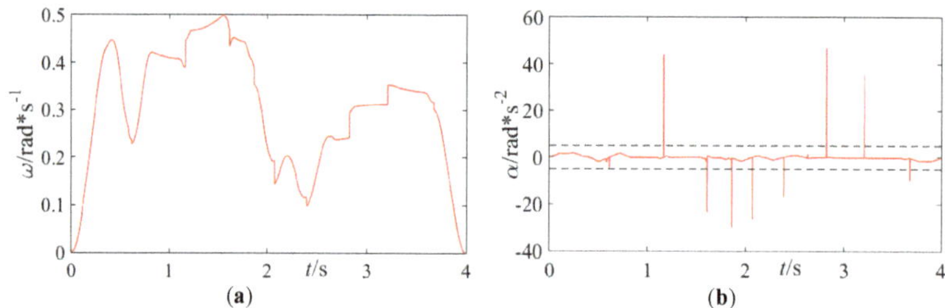

Figure 15. Tool orientation kinematic profiles using parameter synchronization—(**a**) Feedrate profile. (**b**) Tangential acceleration profile.

Since the interpolation results of the tool-tip position and tool orientation are obtained, the reference axial positions are decided according to inverse kinematic transformation from Equation (25). Then the reference axial velocities and accelerations are calculated by differentiating the axial positions. The axial velocities and accelerations does not consider axial constraints are shown in Figure 16. It can be seen that despite a slight improvement of feed time (3.595 s), the acceleration of X-axis and velocity of C-axis exceed the corresponding constraints. The axial velocities and accelerations using the parameter synchronization [23] are demonstrated in Figure 17. The accelerations of all axes are far greater than the allowable values at certain locations. The axial velocities and accelerations using the proposed method is depicted in Figure 18. The data demonstrate that the axial velocities and accelerations are both under limitation. The maximum absolute axial velocities and accelerations using different methods are listed in Table 2 as an intuitive comparison and the anomalously great values are highlighted in bold, especially for axial accelerations using the parameter synchronization method. As can be seen, the large variations of axial accelerations of the traditional parameter synchronization method come from the wild fluctuation in the tool orientation acceleration as shown in Figure 15b.

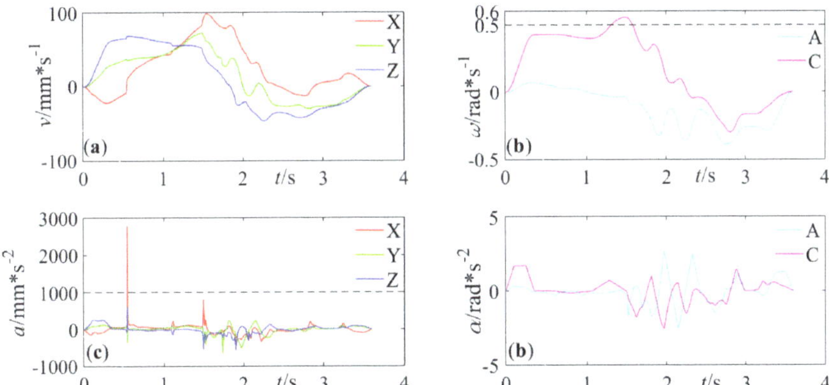

Figure 16. Axial velocities and accelerations that does not consider axial kinematic constraints— (**a**) Translational axial velocities. (**b**) Rotary axial velocities. (**c**) Translational axial accelerations. (**d**) Rotary axial accelerations.

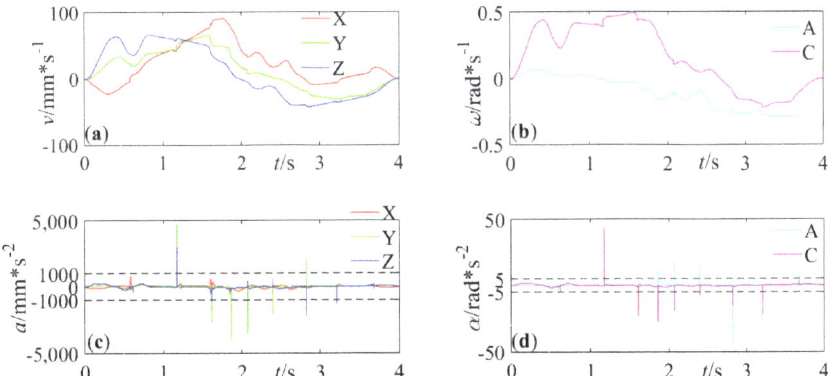

Figure 17. Axial velocities and accelerations using parameter synchronization—(**a**) Translational axial velocities. (**b**) Rotary axial velocities. (**c**) Translational axial accelerations. (**d**) Rotary axial accelerations.

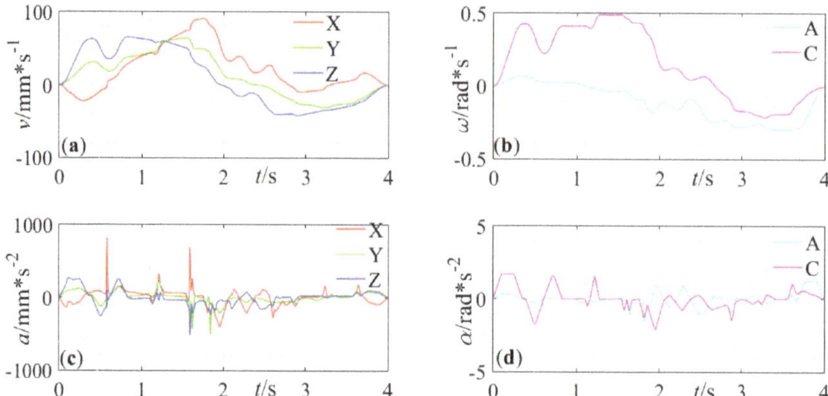

Figure 18. Proposed axial velocities and accelerations—(**a**) Translational axial velocities. (**b**) Rotary axial velocities. (**c**) Translational axial accelerations. (**d**) Rotary axial accelerations.

Table 2. Maximum axial velocities and accelerationsusing different methods.

Methods	Kinematics	X	Y	Z	A	C
Without constraints	Velocities(mm ǀ rad/s)	98.9	72.7	68.6	0.39	**0.56**
	Accelerations(mm ǀ rad/s^2)	**2795**	607.3	569.3	2.67	2.45
Parameter synchronization	Velocities(mm ǀ rad/s)	91.2	65.6	66.4	0.29	0.5
	Accelerations(mm ǀ rad/s^2)	**2871**	**4781**	**2929**	**44.4**	**44.1**
Proposed	Velocities(mm ǀ rad/s)	91.2	64.6	66	0.29	0.49
	Accelerations(mm ǀ rad/s^2)	828.5	493.4	513.4	1.62	1.71

The S-shape test piece made of Aluminum Alloy 1060 is machined on the experimental five-axis table-tilting machine tool with a cutter of diameter 10 mm. The machining process consists of three steps: rough machining, semi-finishing and finishing. The CNC system reads the trajectories using the proposed curve-fitting and feedrate scheduling method and then the axial commands are generated and sent to the controller. The machining results are shown in Figure 19.

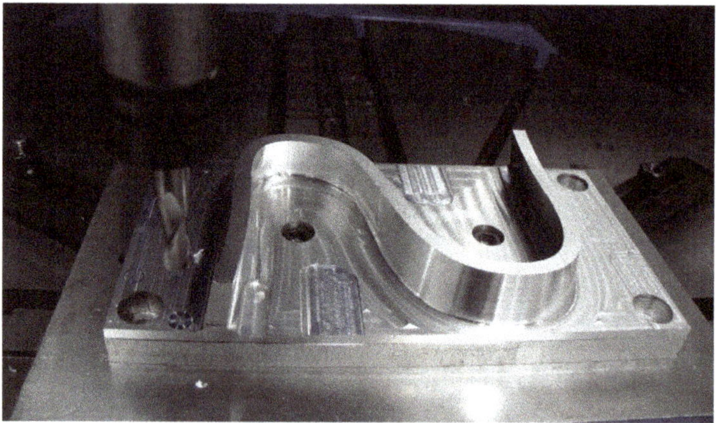

Figure 19. Machining results of S-shaped workpiece.

The machining accuracy of the S-shape workpiece is affected by many factors, such as theoretical error, geometric error of the machine tool, and quality of toolpath [36]. Also, the measurement of freeform workpiece is very complicated. In this experiment, quality of toolpath (fitting error) is mainly concerned. According to drive commands of the machine tool, the real tool-tip position and tool orientation are reversely computed using forward kinematic transformation and the real fitting error in the WCS is calculated. The real maximum and average fitting errors with the traditional parameter synchronization method and the proposed method are demonstrated in Table 3. Due to large variance of axial kinematics, the maximum and average fitting errors with the traditional parameter synchronization show greater values, especially for the tool orientation (40.0% greater for the maximum fitting error).

Table 3. Fitting error of the experiments.

Methods	Position Error (mm)		Orientation Error (deg)	
	Max.	Average	Max.	Average
Parameter synchronization	0.0525	0.0268	0.0717	0.0378
Proposed	0.0507	0.0243	0.0512	0.0317

The real axial velocities and accelerations shown in Figure 20 are obtained by differentiating the real axial positions with the sampling period of 1 ms. The real axial velocities and accelerations are both under control and the large fluctuations are further reduced in the real machining process (the acceleration for X-axis is reduced by 44.3%). Experimental results show that the proposed method can generate trajectories with allowable fitting error and smooth motion.

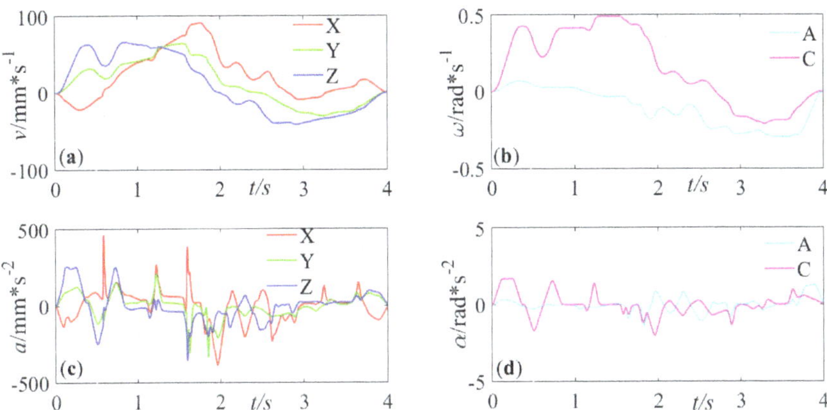

Figure 20. Real axial kinematics—(**a**) Translational axial velocities. (**b**) Rotary axial velocities. (**c**) Translational axial accelerations. (**d**) Rotary axial accelerations.

5. Conclusions

This paper proposes a double B-spline curve-fitting and synchronization-integrated feedrate scheduling method for five-axis linear-segment toolpath. The main technical characteristics are as follows—(1) the tool-tip position and tool orientation are fitted by two cubic B-spline curves respectively so that the discontinuities caused by transition corners of five-axis linear-segment toolpath are handled and the fitting error is strictly respected; (2) the axial kinematic constraints are mapped and satisfied in the feedrate scheduling process of the tool-tip position; (3) the synchronization and smooth rotary motion is obtained using feedrate deformation for tool orientation.

Simulations and experiments are carried out to validate the feasibility and effectiveness of the proposed method. For a toolpath of 12 CL data points from the S-shape test piece, the proposed curve-fitting method constrains the fitting error under fitting tolerance. The proposed machining time is reduced significantly by 44.2% in comparison with the original linear-segment toolpath. To verify the smooth motion of the proposed feedrate scheduling method, we compare the proposed method with the method does not consider axial kinematic constraints and the method using parameter synchronization. The results show that the proposed rotary motion is smooth and the axial motions are under kinematic limits. The experiments are conducted on the S-shape test piece. The real tool-tip position and tool orientation are reversely computed using the real axial positions to confirm that the proposed method can generate a smooth toolpath with allowable fitting error. The real axial kinematics are further smoothed in the real machining process.

Despite the progress made in this study, only fitting error and smoothness are considered for the fitted curve. However, the machining accuracy is affected by other factors such as thermal error, geometric errors of machine tool, dynamic error and so forth, and workpiece measurement to evaluate the surface quality directly is not conducted. Therefore, further research should be done in our future study.

Author Contributions: Conceptualization, X.G.; Formal analysis, S.Z., L.Q., X.L., Z.W. and Y.W.; Investigation, S.Z., L.Q., X.L., Z.W. and Y.W.; Methodology, X.G.; Project administration, S.Z. and L.Q.; Software, X.G.; Writing—original draft, X.G. All authors have read and agreed to the published version of the manuscript.

Funding: This research was funded by the National Natural Science Foundation of China grant number 51675478 and 51875516, and the Natural Science Foundation of Zhejiang Province grant number LY18E050001.

Conflicts of Interest: The authors declare no conflict of interest.

Appendix A. Cutter Location Data

Table A1. Experimental cutter location data.

No.	p_x (mm)	p_y (mm)	p_z (mm)	o_i	o_j	o_k
1	113.560775	7.735266	−2.209314	−0.107258	0.624902	0.773300
2	117.864949	−10.950074	−0.974065	−0.003002	0.653008	0.757345
3	110.502860	−34.808781	1.779567	0.135034	0.648777	0.748902
4	104.086026	−55.840098	2.682644	0.263922	0.596758	0.757776
5	95.225652	−63.959571	4.073524	0.317154	0.551822	0.771301
6	88.820589	−65.972318	6.021818	0.329520	0.516103	0.790603
7	80.162816	−65.096397	7.031464	0.326785	0.478040	0.815285
8	72.478246	−61.109537	6.863103	0.313696	0.446069	0.838223
9	65.985754	−54.666365	6.143037	0.288698	0.416448	0.862105
10	54.251993	−39.568096	4.931174	0.217010	0.357492	0.908354
11	38.038952	−23.111532	3.536073	0.129479	0.261821	0.956391
12	31.679054	−18.711329	3.017623	0.105499	0.220895	0.969575

References

1. Jywe, W.-Y.; Hsieh, T.-H.; Chen, P.-Y.; Wang, M.-S. An Online Simultaneous Measurement of the Dual-Axis Straightness Error for Machine Tools. *Appl. Sci.* **2018**, *8*, 2130. [CrossRef]
2. Wei, X.; Su, Z.; Yang, X.; Lv, Z.; Yang, Z.; Zhang, H.; Li, X.; Fang, F. A Novel Method for the Measurement of Geometric Errors in the Linear Motion of CNC Machine Tools. *Appl. Sci.* **2019**, *9*, 3357. [CrossRef]
3. Zhang, Y.; Ye, P.; Zhang, H.; Zhao, M. A Local and Analytical Curvature-Smooth Method with Jerk-Continuous Feedrate Scheduling along Linear Toolpath. *Int. J. Precis. Eng. Manuf.* **2018**, *19*, 1529–1538. [CrossRef]
4. Liang, F.; Zhao, J.; Ji, S.; Fan, C.; Zhang, B. A novel knot selection method for the error-bounded B-spline curve fitting of sampling points in the measuring process. *Meas. Sci. Technol.* **2017**, *28*, 065015. [CrossRef]
5. Sun, S.; Yu, D.; Wang, C.; Xie, C. A smooth tool path generation and real-time interpolation algorithm based on B-spline curves. *Adv. Mech. Eng.* **2018**, *10*, 1687814017750281. [CrossRef]
6. Zhao, H.; Zhu, L.; Ding, H. A real-time look-ahead interpolation methodology with curvature-continuous B-spline transition scheme for CNC machining of short line segments. *Int. J. Mach. Tools Manuf.* **2013**, *65*, 88–98. [CrossRef]

7. Zhang, Y.; Zhao, M.; Ye, P.; Zhang, H. A G4 continuous B-spline transition algorithm for CNC machining with jerk-smooth feedrate scheduling along linear segments. *Comput. Aided Des.* **2019**, *115*, 231–243. [CrossRef]

8. Fan, W.; Lee, C.-H.; Chen, J.-H. A realtime curvature-smooth interpolation scheme and motion planning for CNC machining of short line segments. *Int. J. Mach. Tools Manuf.* **2015**, *96*, 27–46. [CrossRef]

9. Fan, W.; Lee, C.; Chen, J.; Xiao, Y. Real-time Bezier interpolation satisfying chord error constraint for CNC tool path. *Sci. China Technol. Sci.* **2016**, *59*, 203–213. [CrossRef]

10. Fan, W.; Ji, J.; Wu, P.; Wu, D.; Chen, H. Modeling and simulation of trajectory smoothing and feedrate scheduling for vibration-damping CNC machining. *Simul. Model. Pract. Theory* **2020**, *99*, 102028. [CrossRef]

11. Walton, D.J.; Meek, D.S. G2 blends of linear segments with cubics and Pythagorean-hodograph quintics. *Int. J. Comput. Math.* **2009**, *86*, 1498–1511. [CrossRef]

12. Shi, J.; Bi, Q.; Wang, Y.; Liu, G. Development of Real-time Look-ahead Methodology Based on Quintic PH Curve with G(2) Continuity for High-speed Machining. *Intell. Mater. Mechatronics* **2014**, *464*, 258. [CrossRef]

13. Bi, Q.; Huang, J.; Lu, Y.; Zhu, L.; Ding, H. A general, fast and robust B-spline fitting scheme for micro-line tool path under chord error constraint. *Sci. China Technol. Sci.* **2019**, *62*, 321–332. [CrossRef]

14. He, S.; Ou, D.; Yann, C.; Lee, C.-H. A chord error conforming tool path B-spline fitting method for NC machining based on energy minimization and LSPIA. *J. Comput. Des. Eng.* **2015**, *2*, 218–232. [CrossRef]

15. Lin, F.; Shen, L.-Y.; Yuan, C.-M.; Mi, Z. Certified space curve fitting and trajectory planning for CNC machining with cubic B-splines. *Comput. Aided Des.* **2019**, *106*, 13–29. [CrossRef]

16. Zhong, G.; Liu, P.; Mei, X.; Wang, Y.; Xu, F.; Yang, S. Design Optimization Approach of a Large-Scale Moving Framework for a Large 5-Axis Machining Center. *Appl. Sci.* **2018**, *8*, 1598. [CrossRef]

17. Li, C.; Liu, X.; Li, R.; Wu, S.; Song, H. Geometric Error Identification and Analysis of Rotary Axes on Five-Axis Machine Tool Based on Precision Balls. *Appl. Sci.* **2020**, *10*, 100. [CrossRef]

18. Zhang, J.; Zhang, L.; Zhang, K.; Mao, J. Double NURBS trajectory generation and synchronous interpolation for five-axis machining based on dual quaternion algorithm. *Int. J. Adv. Manuf. Technol.* **2016**, *83*, 2015–2025. [CrossRef]

19. Tulsyan, S.; Altintas, Y. Local toolpath smoothing for five-axis machine tools. *Int. J. Mach. Tools Manuf.* **2015**, *96*, 15–26. [CrossRef]

20. Yang, J.; Yuen, A. An analytical local corner smoothing algorithm for five-axis CNC machining. *Int. J. Mach. Tools Manuf.* **2017**, *123*, 22–35. [CrossRef]

21. Jin, Y.; Bi, Q.; Wang, Y. Dual-Bezier path smoothing and interpolation for five-axis linear tool path in workpiece coordinate system. *Adv. Mech. Eng.* **2015**, *7*. [CrossRef]

22. Yuen, A.; Zhang, K.; Altintas, Y. Smooth trajectory generation for five-axis machine tools. *Int. J. Mach. Tools Manuf.* **2013**, *71*, 11–19. [CrossRef]

23. Li, D.; Zhang, W.; Zhou, W.; Shang, T.; Fleischer, J. Dual NURBS Path Smoothing for 5-Axis Linear Path of Flank Milling. *Int. J. Precis. Eng. Manuf.* **2018**, *19*, 1811–1820. [CrossRef]

24. Wang, Y.; Ma, X.; Chen, L.; Han, Z. Realization methodology of a 5-axis spline interpolator in an open CNC system. *Chin. J. Aeronaut.* **2007**, *20*, 362–369. [CrossRef]

25. Du, X.; Huang, J.; Zhun, L.-M. A complete S-shape feed rate scheduling approach for NURBS interpolator. *J. Comput. Des. Eng.* **2015**, *2*, 206–217. [CrossRef]

26. Lee, A.-C.; Lin, M.-T.; Pan, Y.-R.; Lin, W.-Y. The feedrate scheduling of NURBS interpolator for CNC machine tools. *Comput. Aided Des.* **2011**, *43*, 612–628. [CrossRef]

27. Zhao, D.; Guo, H. A Trajectory Planning Method for Polishing Optical Elements Based on a Non-Uniform Rational B-Spline Curve. *Appl. Sci.* **2018**, *8*, 1355. [CrossRef]

28. Beudaert, X.; Lavernhe, S.; Tournier, C. Feedrate interpolation with axis jerk constraints on 5-axis NURBS and G1 tool path. *Int. J. Mach. Tools Manuf.* **2012**, *57*, 73–82. [CrossRef]

29. Fan, W.; Gao, X.-S.; Lee, C.-H.; Zhang, K.; Zhang, Q. Time-optimal interpolation for five-axis CNC machining along parametric tool path based on linear programming. *Int. J. Adv. Manuf. Technol.* **2013**, *69*, 1373–1388. [CrossRef]

30. Liu, H.; Liu, Q.; Sun, P.; Liu, Q.; Yuan, S. The optimal feedrate planning on five-axis parametric tool path with geometric and kinematic constraints for CNC machine tools. *Int. J. Prod. Res.* **2017**, *55*, 3715–3731. [CrossRef]

31. Erkorkmaz, K.; Chen, Q.-G.; Zhao, M.-Y.; Beudaert, X.; Gao, X.-S. Linear programming and windowing based feedrate optimization for spline toolpaths. *Cirp Ann. Manuf. Technol.* **2017**, *66*, 393–396. [CrossRef]

32. Huang, J.; Lu, Y.; Zhu, L.-M. Real-time feedrate scheduling for five-axis machining by simultaneously planning linear and angular trajectories. *Int. J. Mach. Tools Manuf.* **2018**, *135*, 78–96. [CrossRef]
33. Piegl, L.; Tiller, W. *The NURBS Book*, 2nd ed.; Springer: Berlin/Heidelberg, Germany, 1997; p. 646.
34. Fleisig, R.V.; Spence, A.D. A constant feed and reduced angular acceleration interpolation algorithm for multi-axis machining. *Comput. Aided Des.* **2001**, *33*, 1–15. [CrossRef]
35. Guan, L.; Mo, J.; Fu, M.; Wang, L. Theoretical error compensation when measuring an S-shaped test piece. *Int. J. Adv. Manuf. Technol.* **2017**, *93*, 2975–2984. [CrossRef]
36. Wang, Q.; Wu, C.; Fan, J.; Xie, G.; Wang, L. A novel causation analysis method of machining defects for five-axis machine tools based on error spatial morphology of S-shaped test piece. *Int. J. Adv. Manuf. Technol.* **2019**, *103*, 3529–3556. [CrossRef]

 *applied sciences*

Article

Development of an Analyzing and Tuning Methodology for the CNC Parameters Based on Machining Performance

Ben-Fong Yu * and Jenq-Shyong Chen

Department of Mechanical Engineering, National Chung Hsing University, 40227 Taichung, Taiwan; MichaelChen@dragon.nchu.edu.tw
* Correspondence: headway987@gmail.com; Tel.: +886-4-2284-0165 (ext. 4032)

Received: 14 March 2020; Accepted: 11 April 2020; Published: 14 April 2020

Abstract: This paper proposes the development of a tuning methodology which can set the proper values of the Computer Numerical Control (CNC) parameters to achieve the required machining performance. For the conventional operators of machine tools, the CNC parameters were hard to be adjusted to optimal settings, which was a complicated and time-consuming task. To save time in finding optimal CNC parameters, the objective of this research was to develop a practical methodology to tune the CNC parameters effectively for easy implementation in the commercial CNC controller. Firstly, the effect of the CNC parameters in the CNC controller on the tool-path planning was analyzed via experiments. The machining performance was defined in the high-speed (HS) mode, the high-accuracy (HP) mode, and the high-surface-quality (HQ) mode, according to the dynamic errors of several specified paths. Due to the CNC parameters that have a particularly critical effect on the dynamic errors, the relationship between the CNC parameters and the dynamic errors was validated by the measured data. Finally, the tuning procedure defined the anticipated dynamic errors for the three machining modes with the actual machine. The CNC parameters will correspond with anticipated dynamics errors based on several specified paths. The experimental results showed that the HS mode was the fastest to complete the path, and the completion time of the HP and HQ modes were increased by 37% and 6%, respectively. The HP mode had the smallest dynamic errors than other modes, and the dynamic errors of the HS and HQ modes are increased by 66% and 16%. In the HQ mode, the motion oscillation was reduce significantly, and the tracking error of the HS and HP modes were increased by 85% and 28%. The advantage of the methodology is that it simplifies set-up steps of the CNC parameters, making it suitable for practical machine applications.

Keywords: CNC parameters; machining mode; high speed; high accuracy; high surface quality

1. Introduction

There are many kinds of products produced with Computer Numerical Control (CNC) machine tools. According to the workpiece weight, geometric shape, material, tools, and cutting conditions, the operator needs to estimate the proper manufacturing processes, so the machining requirements will not be the same. Some products require an extremely smooth surface, such as plastic injection molds, while others require high geometrical accuracy and little tolerance, such as precision machine elements (cf., gears), and still others require high cutting efficiency (i.e., material removal rate) and allow larger tolerance, such as the aluminum brackets for bicycle or aerospace. This situation is why machine tools have focused on how to accomplish high-surface-quality (HQ), high-accuracy (HP) and high-speed (HS) machining. Most of the commercial CNC machine tools are equipped with a standard set of CNC parameters, which are not changed according to the different machining demands. The standard set of CNC parameters of machine tools are created by their manufacturers. However, the tuning of CNC parameters has to require a lot of domain knowledge, including the most basic theories, such as statics, dynamics, kinematics, mechanism, motion control, etc. More advanced knowledge also includes the quality of Computer Aided Manufacturing (CAM), interpolation, machine performance, tool geometry, cutting conditions, etc. That is why the machine operators cannot easily tune the CNC parameters to match their product for relevant machining requirements. Figure 1 shows the architecture of feed motion processing sequence in the commercial CNC machine tools in this study. Firstly, the CAM model generates the corresponding Numerical Control (NC) code. The position command of the NC code (S_g) would be interpreted by the CNC controller for tool-path planning in interpolator, and that generates the motion trajectory based on the CNC parameters. Therefore, the CNC parameters include the velocity, acceleration, and jerk for their path statements, such as straight, corner, and arc paths. An algorithm on the tool-path planning produced the kinematic profiles of their relevant position, velocity, acceleration, and jerk based on NC code and CNC parameters. According to the kinematic profiles, each servo axis would move along the trajectory motion that was defined as position command (S_c). The S_c drives the feed drive system of the machine tools via servo loop. The feedback signal was defined as the actual position (S_a) for the feed drive system in the machine tools. Finally, the real workpiece would be produced based on the tool path, tool, material, and cutting conditions. The loop of the blue arrow in Figure 1 has been widely used in all machine tools.

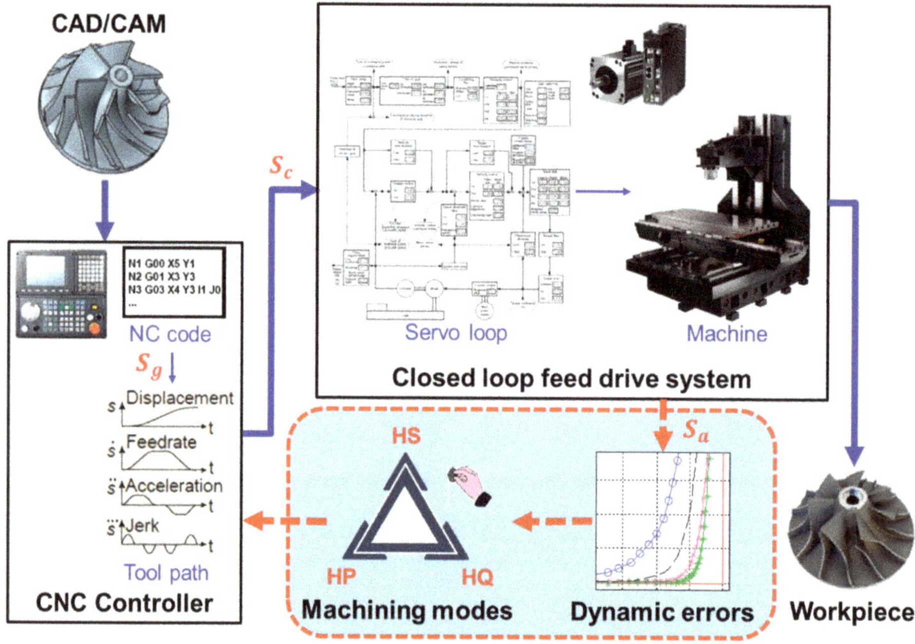

Figure 1. The goal architecture of the processing sequence in this study.

The goal of this study was to analyze the relationships among machining modes, CNC parameters, and dynamic errors, followed by developing the tuning procedure which can tune the proper values of the CNC parameters to match the selected machining mode; it is shown on the red dot frame in Figure 1. Firstly, an analysis of the CNC parameters based on algorithm of tool-path planning in the conventional controller, and the principle of tool-path planning was organized. It proposed the dynamic errors chain that we have to observe in actual machine tools. In this paper, dynamic errors signify the deviation between the position of the NC code and the actual position of the feed drive system. It contains two parts: One part is the difference between the S_g and S_c, another part is the difference between the S_c and S_a. Secondly, we experimented on the machining modes with different CNC parameters and collected data. It presents a tuning criteria for the three machining modes and CNC parameters. Finally, this study adjusted the CNC parameters and analyzed the experimental results with an actual machine to develop the tuning methodology for machining modes. It proposes the tuning methodology and procedure for the CNC parameters based on machining requirements.

A number of studies have analyzed the effects of the tool-path planning to follow the desired contours, and jerk control strategies have been proposed [1–4]. According to the path distance and feedrate of the NC code, the form of the acceleration profile must first be defined by the presetting of jerk and acceleration for CNC parameters; the acceleration profile of the triangle or trapezoid was planned. Some literature introduced the advanced method for interpolation, such as Non-Uniform Rational B-Splines (NURBS) and Bazier [5,6]. Sencer et al. [7] literature presented the algorithm with learning for NC code re-established. Therefore, the objective of the tool-path planning is to ensure the feedrate can be smoothed for precision results as required. Tapie et al. [8] proposed the circular tests for high speed machining, and proposed the limit values of jerk, acceleration, and circular velocity. However, the tool path is not only a circular path, but also contains straight and corner paths. This study did not discuss the other CNC parameters. Pateloup et al. [9] presented the corner optimization. However, it proposed the corner with arc path for pocket machining. In general, the corner path consists of two short lines, and the limit of corner velocity avoids the vibration after the corner motion. This study did not obtain the corner velocity of CNC parameters.

Andolfatto et al. [10] presented the dynamic errors at the tool center point, and included the contour error caused by a closed loop feed drive system and actual error from the machine structure that defined the feedback position to relate tool center point. However, this study did not discuss the error caused by CNC parameters. Lee et al. [11] also established a servo parameter tuning method based on the corner error. The experimental results based on the corner error measurements showed that the proposed tuning methods are very effective in improving the corner accuracy of a three-axis high-speed CNC machine tool. Parenti et al. [12] proposed the acceleration spectrum of the acceleration profiles which exert significant effects on the behavior of the dynamic errors in the CNC machine tools at the tool center point. Bringmann et al. [13] introduced a measurement device based on the "R-test" for measuring displacements in the three linear degrees of freedom, and for evaluating the dynamic errors of the machine at the tool center point. The CNC parameters, such as jerk and acceleration limits, can be set to obtain the required dynamic path accuracy. However, the proposed tuning method was proven only for the HP mode, and not for the HQ or HS mode. Li et al. [14] presented the different feedrate effect on the dynamic errors of the feed drive system. This proposal was useful for HQ mode, but this study did not have any solutions to tune the other modes.

Some studies have analyzed the influence of motion errors for the machining performance. Some scholars focus on the dynamic errors caused by tool geometrics [15–17]. It is dependent on a larger machine and the dynamic errors' source on machine structure. Mia et al. [18] presented intelligent optimization of hard-turning parameters. This paper proposed the evolutionary algorithms for cutting conditions. Their dynamic errors adjust method was to change the cutting conditions. The main disadvantage is that the error between the measuring tool center point and the geometry of NC code must be considered. This is not a trivial task for the machine's operator. Li et al. [19,20] introduced the ambient environmental temperature effect on the positioning accuracy of moving carrier on machine tools. Therefore, the carrier embedded cooling channels to achieve thermal error suppression, and the machine tools were operated in an air conditioning chamber in which the temperature could be controlled. However, the variation of accuracy caused by thermal deformation of machine tools comes from the long-term machining. Actually, there are not any CNC parameters to solve the problem of thermal displacement.

Several works developed the virtual machine tool technology which predicts the contour error of the closed loop feed drive system of the CNC machine tools [21–26]. However, the virtual machine tool technology proposed by these previous works involved a mechatronic model which integrated the servo driver, mechanics of the feed drive mechanism, and the dynamics of the machine structures. Although the effects on the dynamic errors after changing the CNC and servo parameters have been studied, they did not establish or propose any tuning algorithm to switch the machining performance between HS, HP, and HQ modes.

Conventional controllers have been provided with the functions of the HS, HP, and HQ modes in the CNC machine tools; they are predominantly used in trajectory generation [27,28]. When the user selects a machining mode, the controller has to process feedrate with the NC blocks. In order to achieve the machining strategy, the control method has to match the tolerance value of the path contour via a relevant NC program. However, the algorithm of different machining modes isn't included in the feed drive system of real CNC machine tools; the function is merely processed in the interpolation of the controller.

This paper presents a tuning methodology for the CNC parameters, and, hence, matches the machine's dynamic errors according to the selection of machining modes which are set by the mixed combination of three machining modes: The HS mode, HP mode, and HQ mode. The relationship between CNC parameters and dynamic errors is defined in Section 2. Section 3 uses simple paths to test the tuning criteria of the CNC parameters and different machining requirements. The tuning methodology and procedure of the CNC parameters based on machining modes were implemented in this section. In Section 5, the CNC parameters were estimated based on the relevant machining modes for experimental validation. Finally, the conclusion is presented in Section 6.

2. The Definition of the CNC Parameters and Dynamic Errors

2.1. Tool-Path Planning in the CNC Controller

The function of the interpolator is to generate the velocity profile of the tool path, so the interpolator needs to provide smooth velocity change during the contour machining via a multi-axis machine tool. The tool-path planning algorithm maintains a smooth velocity transition during the high-speed feed motion. Most commercial CNC controllers use S-curve velocity profile along with the function of jerk control. According to the specific feedrate and displacement from the NC code, the kinematic profiles (velocity) are generated by the interpolator in the CNC controller with the CNC parameters (jerk, acceleration, and time constant of the acceleration/deceleration (Acc/Dec)).

Figure 2 shows the typical trapezoidal acceleration profile. For feed motion along the tool path, acceleration profiles are linear, velocity profiles are parabolic, and position profiles are cubic for regions 1, 3, 5, and 7 where time constant of Acc/Dec with constant jerk occurs. Acceleration values are constant and jerk is zero for regions 2 and 6, where velocity profiles are linear and position profiles are parabolic. In region 4, jerk and acceleration values are zero, velocity is constant, and position is linear. If the feed motion has the greatest acceleration, the acceleration profile does not have the regions 2 and 6. When the triangular acceleration profile has been generated, the result is no constant acceleration, but it has a short acceleration peak. While this profile is useful to reduce machining time, it has very high requirements on the performance of the machine. Finally, the result of the tool-path planning is divided into the position command of each feed axis to enter the servo loop, and the motor that drives each feed axis reaches the position required for the workpiece shape.

Actually, the tool-path geometry not only consists of a straight path, but also includes circular and curved paths generated from many short lines. The general purpose machine tool consists of three mutually perpendicular linear axes, and the multi-axis synchronous motion makes the tool-path process a complex shape. Figure 3 shows the tool path with complex shape. It consists of many short lines. When the X and Y axes are in synchronous motion for the tool path, the path velocity will be reduced to distribute the X-axis velocity (V_x) and Y-axis velocity (V_y). The P_1 and P_2 are the corner motion. The feedrate must be decreased at t_{P1} and t_{P2} (as shown in Figure 3). One of the CNC parameters is to set a corner velocity (V_c) for the tool-path planning; the V_c is composed of the V_x and

V_y. The purpose is to define the time point at which the V_y starts to increase velocity when the V_x has decelerated to the V_c.

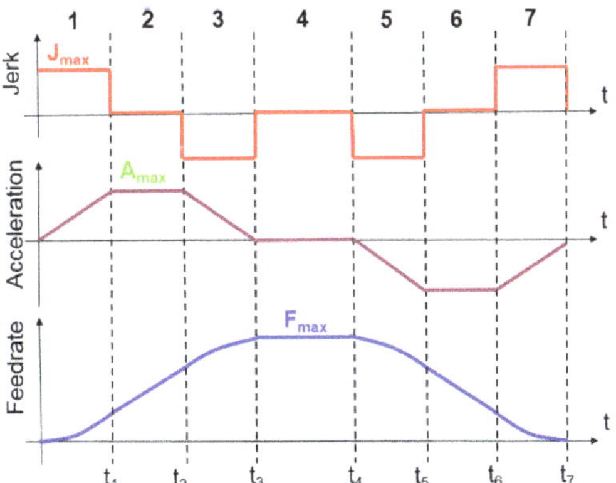

Figure 2. Typical trapezoidal acceleration profile.

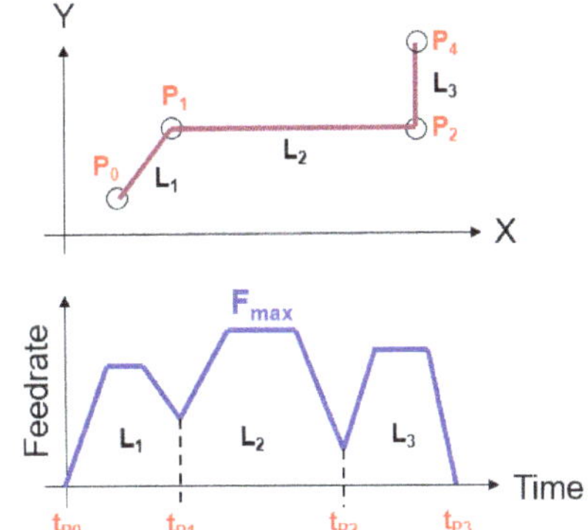

Figure 3. The velocity profile with the complex shape.

The NC code is circular with the radius (R), and the circular path is composed of the X-axis and the Y-axis. In fact, the path velocity (V_p) has to be consistent, while the V_x and V_y are changing constantly, and each axis is constantly accelerating and decelerating at the same time. Tapie et al. [8] showed that the velocity profiles are similar to the sine wave during the feed motion. There are two kinds of CNC parameters to limit the velocity of the circular path for the interpolator in a commercial controller. One is the arc velocity (V_a) and the other is centripetal acceleration (A_c). The feedrate of NC code (V_f) is similar to the V_p, but the V_f is only used in the velocity of the circular path for the tool-path planning. The relationship between the centripetal acceleration and the arc velocity is expressed as:

$$A_p = \left(V_f\right)^2 / R. \tag{1}$$

When V_f is larger than the feed motion, the greater centripetal acceleration is excited, causing the error between the circular path and the ideal path at the same time.

The flow chart of the tool-path planning was organized as shown in Figure 4. According to the path geometry and feedrate of the NC code, the form of the acceleration profile must first be defined. The profile of the triangle or trapezoid was planned by presetting jerk (J_p), acceleration (A_p), and time constant of the Acc/Dec (T_p). These values can be preset by the CNC parameters of the controller. The actual acceleration (A_p') will be defined by the tool-path planning. It includes the corner velocity

(V_c), arc velocity (V_a), and smoothing time constant (T_a) for the velocity profile with path velocity (V_p), and, finally, the completed position command required for each axis (S_{cx}, S_{cy}, S_{cz}). The most important function of the controller's interpolator is making the velocity profiles of each axis as smooth as possible to avoid tool-path drastic speed changes. The drastic speed changes would produce force, causing machine deformation or oscillation during the machining process, which would affect the accuracy of the workpiece and surface. This study adjusted the parameters and analyzed the experimental results with an actual machine to develop the tuning methodology for machining modes.

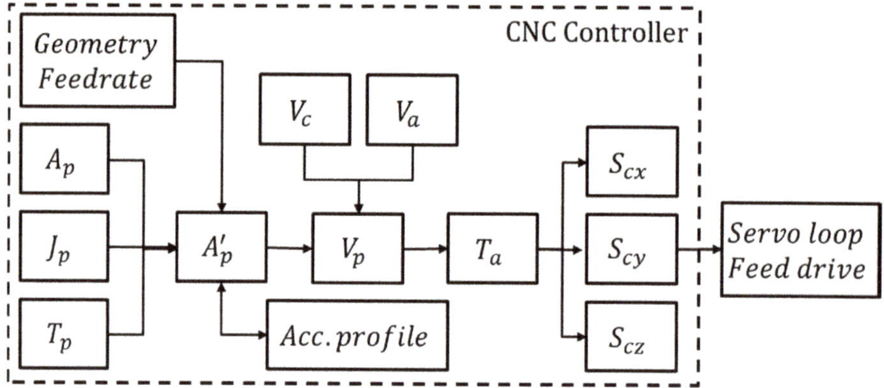

Figure 4. The principle of the tool-path planning in interpolator.

2.2. The Dynamic Errors' Chain Corresponding to the Trajectory Generation

Dynamic errors signify the deviation between the path of the NC code and the actual path. They contain the error from the NC code after interpolation in the controller that is a command error. In Figure 1, the command error (ε_1) is the deviation between the original command from the NC code (S_g) and the position command (S_c) through the interpolator, expressed as:

$$\text{Command error} = \varepsilon_1 = S_g - S_c. \tag{2}$$

The command error represents the influence of the CNC parameters in the tool-path planning.

Contour error (ε_2) is another dynamic error, caused by the dynamic characteristics of the servo loop and feed drive system. The corresponding contour error is the deviation between the position command (S_c) and actual position (S_a) expressed as:

$$\text{Contour error} = \varepsilon_2 = S_c - S_a \tag{3}$$

where S_c and S_a are position command and actual position, respectively. Numerous studies [4,11,17] used the contour error to analyze the performance of machine tools. The phenomenon of servo delay causes position error, called tracking error, on a single axis, and is named as contour error if the path was simultaneously moving by multi-axis. The dynamic errors (ε_3) include ε_1 and ε_2, introduced as:

$$\text{Dynamic errors} = \varepsilon_3 = \varepsilon_1 + \varepsilon_2 = S_g - S_a \tag{4}$$

The ε_1 was caused by the interpolator in tool-path planning, while the ε_2 was caused by the closed loop feed drive system of the machine. Figure 5 shows the dynamic errors chain with ε_1 and ε_2 for the corner path and the circular path, respectively. In the following description, we focused on the development of the HS/HP/HQ mode by observing the dynamic errors during the axis movement.

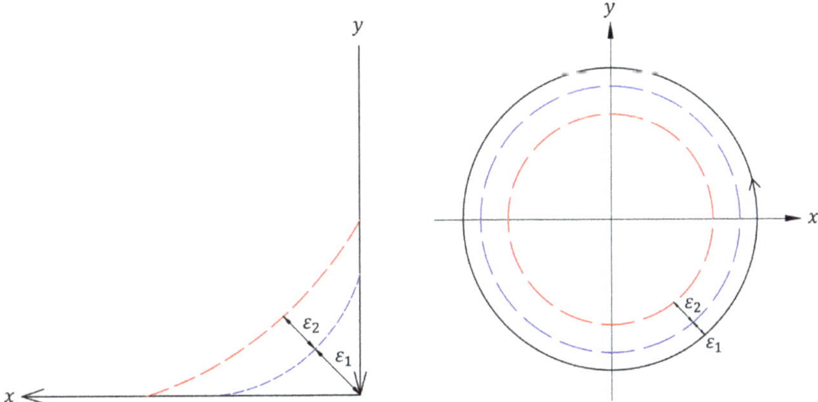

Figure 5. The dynamic errors in corner path and circular paths.

The tool-path planning purpose is to control the path velocity and slow down the velocity change of each axis when the machine moves along the three-dimensional path. When the velocity of each axis changes drastically, the generation of acceleration causes the machine to be stressed and thus generates errors and vibrations.

3. Experiment on the Machining Modes with CNC Parameters

3.1. CNC Parameters' Tuning Criteria for the Machining Modes

Different workpieces have different requirements in the machining process, such as aerospace or difficult-to-cut materials, which require higher efficiency for material removal; the surface quality and accuracy are not the priority requirements, as defined in the HS mode. Regarding the structure of the wing used in aerospace, the product demand is high geometric accuracy. As surface quality was not the priority requirement for this product, it was defined via the HP mode. The mold of the optical component needs to have good surface quality and the accuracy needs at least to match the requirement, so it defines with the HQ mode. The tuning criteria of the CNC parameters for the three kinds of the machining mode are summarized in Figure 6. The relationship between the CNC parameters' tuning criteria of the three machining modes is defined below:

- HS mode: The objective is to set the feed motion of machine tools suitable for the larger dynamic errors. It will finish the tool path with the shortest time for motion. In order to reach the target, the CNC parameters' values of jerk, acceleration, corner velocity, and arc velocity must be increased. At the same time, the HS mode will cause larger oscillation after the Acc/Dec motion, and it will mark the vibration texture on the surface.
- HP mode: The dynamic errors should be kept to a minimally tight tolerance. The motion path is close to the ideal path of the NC code. In order to reach the goal, the values of jerk, acceleration, corner velocity, and arc velocity must be reduced for HP mode. Therefore, the smaller values of the CNC parameters result in larger cycle time.
- HQ mode: The goal is to obtain a smoother machining surface. This means that we must maintain minimal oscillation during feed motion. In addition to reducing the values of jerk, acceleration, corner velocity, and arc velocity, smoothing time constant must be provided to restrain the mechanical resonance. The magnitude of the dynamic errors is between the HS mode and HP mode. In the HQ mode, the motion hardly oscillates, and, therefore, contributes to the smooth surface quality on the geometric surface of the product.

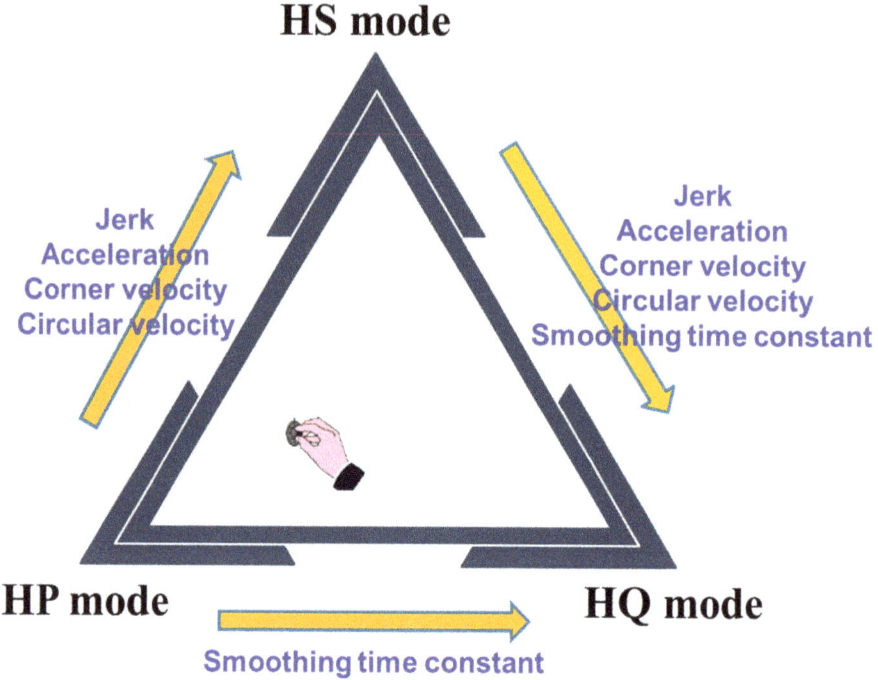

Figure 6. The tuning criteria of the three machining modes.

3.2. Experimental Platform

The experiments were performed on a three-axis machine tool with SYTEC's PC-based CNC controller, and DELTA's ASDA-A2 servo motor and servo driver. The absolute encoder was implemented in the servo motor and the linear scale was mounted on the feed drive system. The resolution of linear scale was 1 μm for the HEIDENHAIN LS100-series. The software of ASDAsoft was used to ensure the operating system with real-time performance.

The experiments presented how to verify the validity of the tuning criteria of the three machining modes based on several specified paths, such as straight, square, and circular paths. According to the requirements of tool-path planning, the relevant CNC parameters are listed in Table 1. In addition to the path geometries and feedrate of NC code, the acceleration profile of the triangle or trapezoid was planned via the CNC parameters in the table for the tool-path planning, such as preset jerk (J_p, N100), acceleration (A_p, N200), and time constant of Acc/Dec (T_p, N300), and then included the corner velocity (V_c, N500), arc velocity (V_a, N600), and smoothing time constant (T_a, N400). The J_p, A_p, and T_p will be limited by the CNC parameters of each axis based on the tool-path planning in the interpolator, such as the J_x (N101), J_y (N102), J_z (N103), A_x (N201), A_y (N202), A_z (N203), T_x (N301), T_y (N302), and T_z (N303). Therefore, when testing for specified paths, the J_x, J_y, J_z, A_x, A_y, and A_z have to be set to extreme limits and the T_x, T_y, and T_z have to be set to zero. Its advantage is that the J_p and A_p of the path can be fully presented. It is useful to estimate the relationship between the CNC parameters and dynamic errors. In addition, in order to understand the efforts of CNC parameters, the feedrate has to be kept the constant value for specified paths in the NC code. The geometrical shape of NC code is S_g as the idea position, the CNC controller completing the position command (S_{cx}, S_{cy}, S_{cz}) required for each axis servo loop, and the feed motion resulted in actual position (S_{ax}, S_{ay}, S_{az}) source from linear scale. The S_{cx}, S_{cy}, S_{cz} and S_{ax}, S_{ay}, S_{az} were measured and stored by the ASDAsoft, and the numerical analysis was performed with MATLAB.

Table 1. CNC parameters corresponding to the tool-path planning.

No.	Symbol	Description	Unit
N100	J_p	Maximum jerk for tool-path movement	m/s^3
N101~203	J_x, J_y, J_z	Maximum jerk for each axis movement	m/s^3
N200	A_p	Maximum acceleration for tool-path movement	m/s^2
N201~203	A_x, A_y, A_z	Maximum acceleration for each axis movement	m/s^2
N300	T_p	Time constant of Acc/Dec for tool-path movement	ms
N301~303	T_x, T_y, T_z	Time constant of Acc/Dec for each axis movement	ms
N400	T_a	Smoothing time constant for tool-path movement	ms
N500	V_c	Maximum corner velocity for tool-path movement	mm/min
N600	V_a	Maximum arc velocity for tool-path movement	mm/min

According to the various paths and the relevant CNC parameters, we designed the experimental conditions to validate how the CNC parameters influence the dynamic errors when the machine was set as HS, HP, and HQ mode individually. On the other hand, the experimental data was collected as an important basis for the tuning methodology of CNC parameters.

3.3. Parameters' Analysis by Straight Path

Figure 7 shows the experimental results of the straight path with distance 350 mm and feedrate 60,000 mm/min. Herein, three selections of CNC parameters were introduced for the HS mode, as shown in Figure 7 (blue solid line: N100 is 200 m/s^3, N200 is 6.8 m/s^2; green dashed line: N100 is 200 m/s^3, N200 is 9.8 m/s^2; red dotted line: N100 is 100 m/s^3, N200 is 9.8 m/s^2) and the command signals for position, velocity, acceleration, and jerk were observed by the CNC controller. With the blue solid line with larger jerk and smaller acceleration limit, it was easy to reach the accelerated limitation. The acceleration profile was trapezoidal and the path completion time was t_{l1}. The green dashed line had the increased acceleration limitation. The acceleration was larger than in the first condition, and the time of Acc/Dec was shorter than in the first condition, so that the path completed time t_{l2} was shortened. With the red dotted line, although the accelerated limitation was larger and the jerk was smaller, the acceleration profile was triangular. The third condition path completed time was the longest t_{l3}. Finally, the path completion time was compared to $t_{l3} > t_{l1} > t_{l2}$, and the t_{l2} was the shortest time to complete the machining process. This result verified that the HS mode had larger J_p and A_p.

The first condition had a short acceleration time, but it was limited by the acceleration setting parameter N200. Therefore, the profile was a trapezoidal acceleration profile. This was a disadvantage for the machine because the machine received force for a long period, which made machine dynamic errors or oscillations happen more easily. The second condition set an upper acceleration limitation, thus shortening the segment time during acceleration and deceleration. If the dynamic characteristics' performance of the servo and the machine are raised, the maximum accelerated limitation can be increased, so that it can be a triangular acceleration profile, and the path completion time can be reduced. The acceleration profile was triangular for the third condition. The jerk could reach the maximum acceleration directly, but the maximum acceleration was not large enough, making the finishing time longer; the path completion time was finally increased. However, the larger J_p and A_p will excite the resonance of the machine, which will cause vibration during the feeding process, and the workpiece surface will suffer worse oscillation textures. Since HQ mode pursues good surface quality, it needs to maintain a constant feedrate by avoiding acceleration and deceleration during the motion, since the feed force instantaneously causes vibration on the machine (as shown in Figure 8) during acceleration and deceleration. Then, some distinct vibration caused the surface of the workpiece to become rough. However, it is impossible that no acceleration or deceleration occurred during the motion of the three-dimensional path. Therefore, the HQ mode is not meant to preclude small dynamic errors. Its significance is in reducing the oscillation caused by acceleration and deceleration of each axis. The oscillation of the Acc/Dec machine cannot be defined by the ε_1 after interpolation; the oscillated behavior resulted from the dynamic characteristics of the real machine. Therefore, the HQ mode was used to observe the tracking error of each axis, reduce the oscillation of the feed drive system during the period of acceleration and deceleration, and avoid the rough surface on the workpiece. The Acc/Dec is equivalent to an impact on the input of the mechanism. If the bandwidth of the impact is greater than the resonance mode of the machine, the resonance is easily excited. Generally, the controller smoothed the oscillation during acceleration and deceleration by the filter for the time constant parameter. The function of the filter was to ensure smooth path velocity but it increased the completion time of the machining and the dynamic errors of the path. Figure 9 shows the experimental results of a straight path whose distance was 25 mm, acceleration was 10 m/s^2, and feedrate was 6000 mm/min with different filter T_a (N400). Figure 9 shows the tracking error for a single axis, and it also shows the spectrum of the fast Fourier transformation (FFT) based on the tracking error. The spectrum can be observed that there were two peaks caused by the oscillation of the feed drive system. Their resonance frequencies were 38 and 46 Hz, respectively. The first resonance should have been suppressed by the proper T_a. It can be observed that after the T_a was 26 ms, the first resonance was suppressed. If the T_a was based on the second resonance frequency, then the oscillation still was excited for the feed drive system. For the tracking error, it was found that after increasing T_a, the oscillation was reduced after passing of the Acc/Dec (red line in Figure 9). In addition, when the T_a was 0, the path completion time was t_{s1}. When the T_a was 26 ms, the path completion time was extended to t_{s2}. In this case, the T_a was 0 ms for the HS mode, and the T_a was 26 ms for the HQ mode.

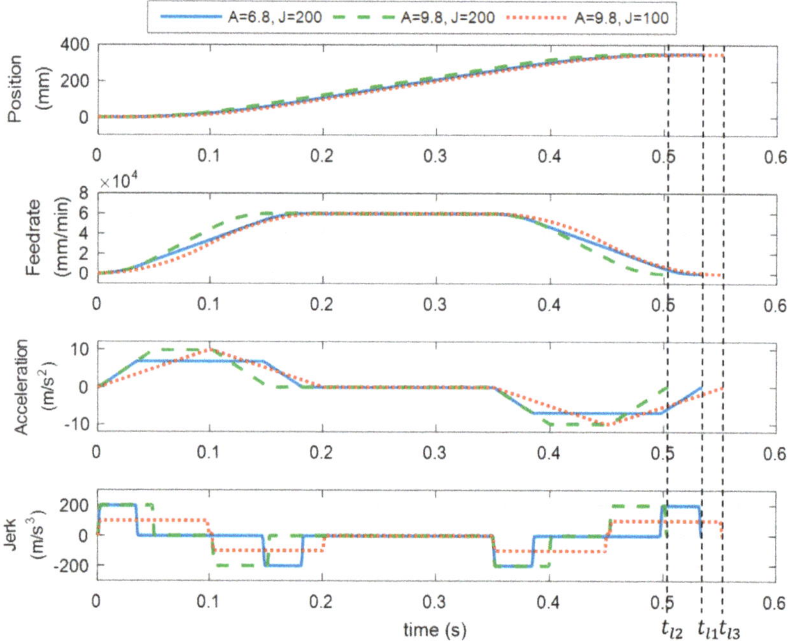

Figure 7. Analysis of straight path with different parameters.

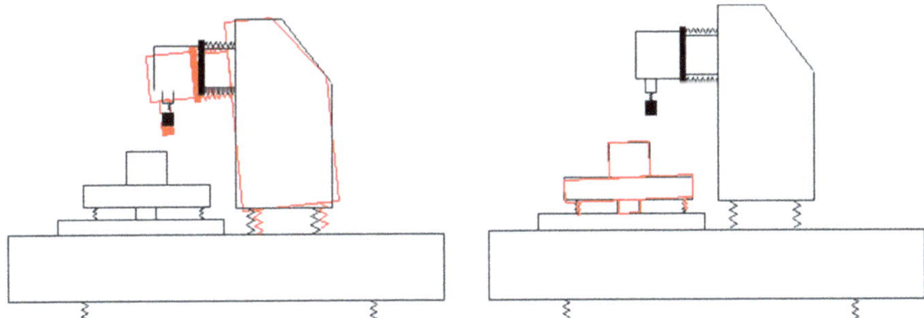

Figure 8. The oscillation caused by feed motion in the feed drive system.

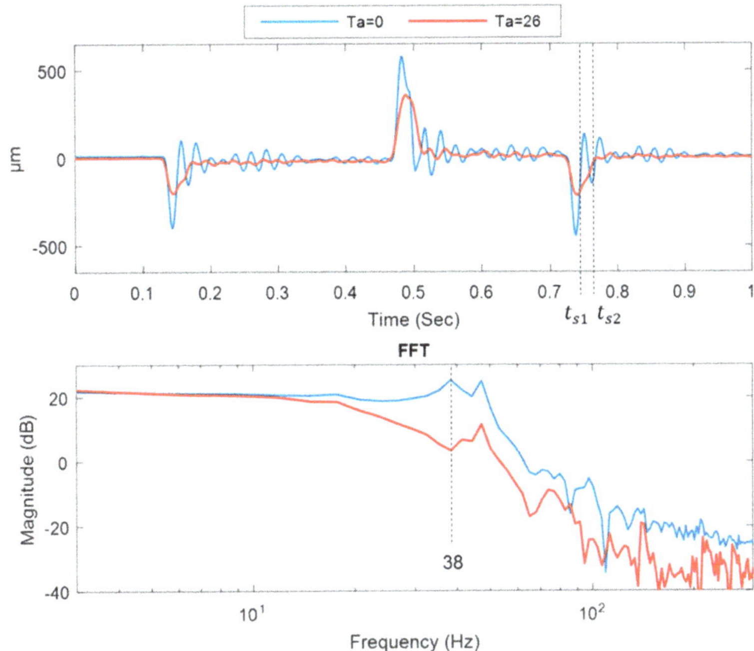

Figure 9. The smoothing time constant (T_a) effect on the tracking error and spectrum.

3.4. Parameters' Analysis by Square Path

Figure 10 shows a square path in the left with a side length of 30 mm, feedrate was 5000 mm/min, acceleration was 4 m/s^2, and jerk was 200 m/s^3 by setting the three different corner velocities (N500), which were 100/1000/3000 mm/min, respectively, and observing the influence of the V_c at the corners for C_1, C_2, and C_3. The right of Figure 10 shows the testing result of the C_1. The larger V_c caused larger dynamic errors (green line in Figure 10), and the period time of the path S_c was shorter. The smaller V_c caused smaller dynamic errors (blue line in Figure 10), and the period time of the path S_c was longer.

Figure 11 shows a velocity profile of a typical square path. The diagrams from top to bottom were the V_x, V_y, and V_p, respectively. The completed time was $t_{c1} > t_{c2} > t_{c3}$ in these results. Figure 12 shows the acceleration profile of the square path. The diagrams from top to bottom were A_x and A_y, respectively. First, observing the velocity and acceleration of C_1, C_2, and C_3, the larger value of the V_c would cause the velocity at the corners to suddenly become discontinuous, resulting not only in a large acceleration but also the A_x and A_y becoming unstable (green line in Figure 12) at the corners;

oscillation would occur in the constant velocity section. The smaller V_c would cause smooth velocity because the A_x and A_y reached the set value of the parameter at the corner. The performance of the HP mode included the dimensional accuracy and geometric accuracy of the workpiece, which means the dynamic errors needed to be smallest when the path was performed. In this case, the V_c was 3000 mm/min for the HS mode, and the V_c was 100 mm/min for the HP mode. However, if the V_c is largest, the completion time is shortest for the path, but the dynamic errors will increase on the corner at the same time.

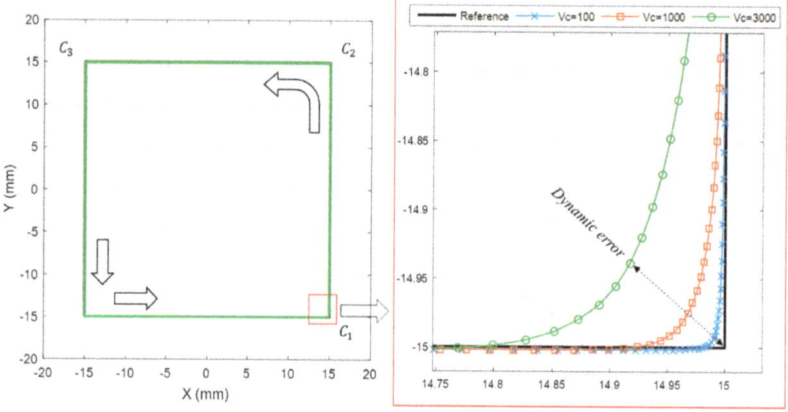

Figure 10. The dynamic errors of the square path.

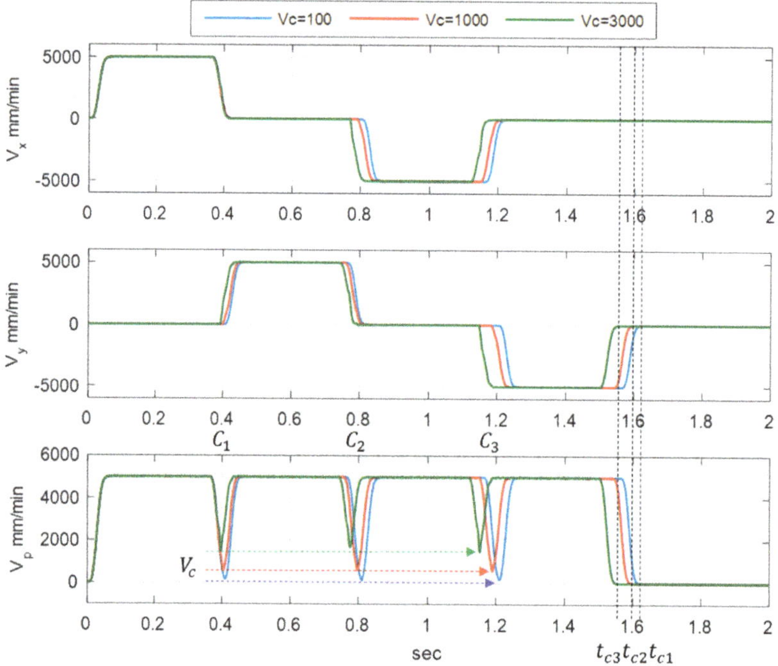

Figure 11. Analysis of square path with different corner velocity (V_c).

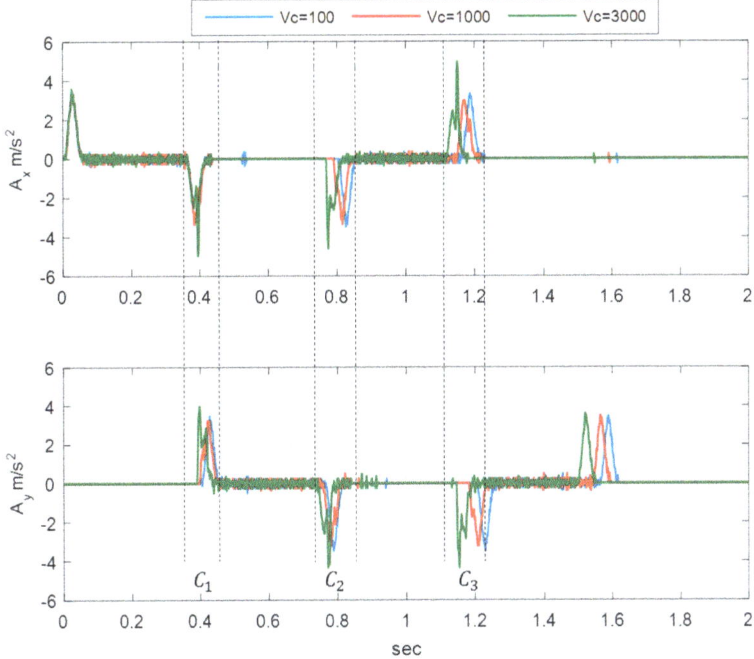

Figure 12. Analysis of the acceleration at each corner with different corner velocity (V_c).

3.5. Parameters' Analysis by Circular Path

Figure 13 shows the experimental results of a circular path with a radius of 15 mm, acceleration of 4 m/s², and jerk of 200 m/s³. Adjust the three conditions of different V_a (N600) (blue line: 3000 mm/min; red line: 6000 mm/min; green line: 9000 mm/min) and observe the feedback signals. In order to easily observe the testing results with variant V_a, the feedrate of NC code must be greater than 9000 mm/min; otherwise, V_a is limited by the feedrate. Figure 14 shows the conversion of a circular path into a straight line. The diagrams from top to bottom show that actual radius value and dynamic errors were due to different V_a, respectively. The horizontal axis in the error graph is the angle, and the black line is the reference path. The four quadrants of the circular path, A_1, A_2, A_3, and A_4, were the reverse directions of the axis, at the position of the axis where acceleration and deceleration occur. The acceleration value was limited by parameters including N101~103, N201~203, and N301~303. The sharp corners caused the backlash of the feed drive system, and the sharp corners' error value reduced the dynamic errors, so that the influence of A_1, A_2, A_3, and A_4 should not be estimated in the dynamic errors. To analyze the influence of V_a on the circular path, a large V_a will produce large dynamic errors, and a smaller V_a will produce smaller dynamic errors.

Figure 15 shows the velocity curve for a typical circular path. The diagrams from top to bottom present V_x, V_y, and V_p. The V_x and V_y are sine waves, and V_p had a constant velocity except at the start position and the end position. Comparing the completion time of the path, the result was $t_{a1} > t_{a2} > t_{a3}$. In this case, the V_a was 9000 mm/min for the HS mode, and the V_a was 3000 mm/min for the HP mode. However, if the V_a is largest, the completion time is shortest for the path, and the dynamic errors increase at the same time.

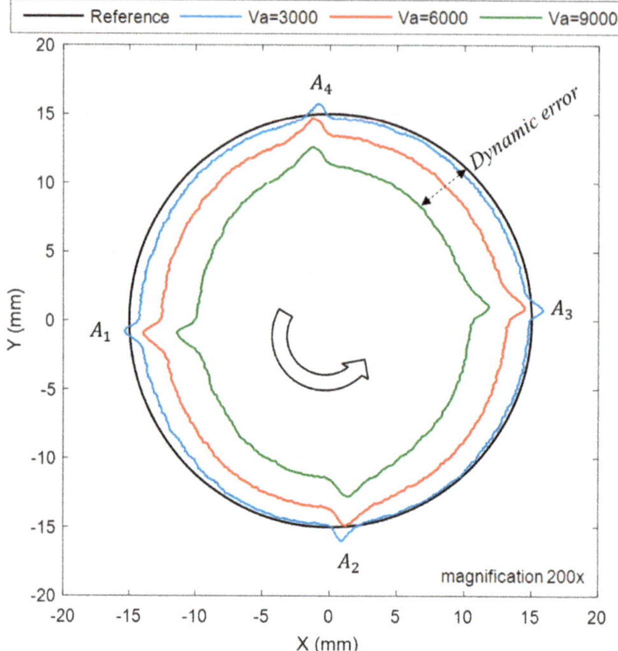

Figure 13. The dynamic errors of the circular path.

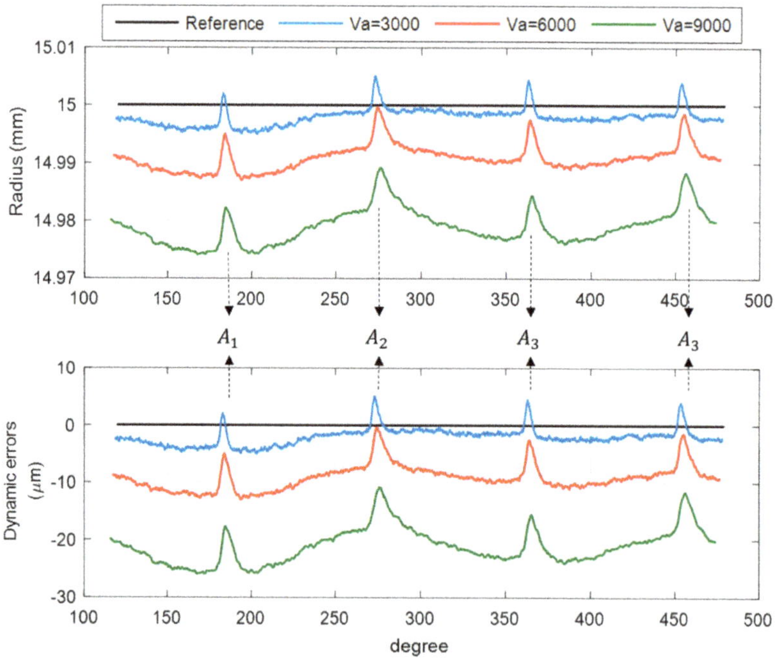

Figure 14. Analysis of circular path for the dynamic errors at all degrees of the whole path.

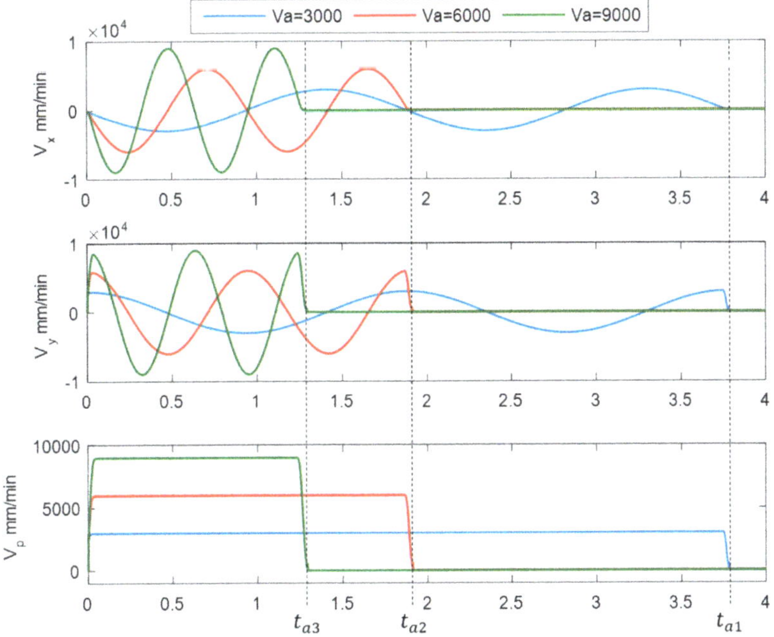

Figure 15. Analysis of circular path for the velocity of the whole path.

In summary, in order to reduce the dynamic errors, the jerk, acceleration, V_c, and V_a need to be reduced properly, so that the advantage of HP mode is fewer dynamic errors, but the path completion time becomes longer. In general, HP mode is used for the finishing process, and most machining requirements are dimensional accuracy and geometric accuracy. However, mold processing needs more dimensional accuracy requirement, not geometric accuracy.

4. Tuning Methodology and Procedure for the CNC Parameters

The relationship between CNC parameters and the three machining modes was verified in Section 3. The experimental results showed the effectiveness of our proposed machining mode for corresponding CNC parameters, which is relevant to the dynamic errors of the machine. The tuning methodology and procedures of the CNC parameters were defined by the simple path, as shown in Figure 16. Firstly, the anticipated dynamic errors were defined as the maximum magnitude of the tool path that had first been allowed. In this paper, the anticipated dynamic errors of the HS/HP/HQ mode were set to 0.05/0.01/0.02 mm, respectively. Actually, the anticipated dynamic errors had no absolute value. They must to be adjusted by the operator for the product requirements. The specified paths used the relevant feedrate for testing. We suggest the proper feedrate that the machining condition used. According to anticipated dynamic errors and the circular test, the A_p and V_a could be easily calculated based on Equation (1), secondly. For the A_p and straight test, the J_p was calculated. The T_a and V_c were estimated based on the square test. The tuning procedure was based on actual measurement of the machine's two-dimensional contouring performance. The machine's dynamic errors were measured with simple paths, and the best combination of CNC parameters were chosen. A practical methodology will be discussed in this section in detail.

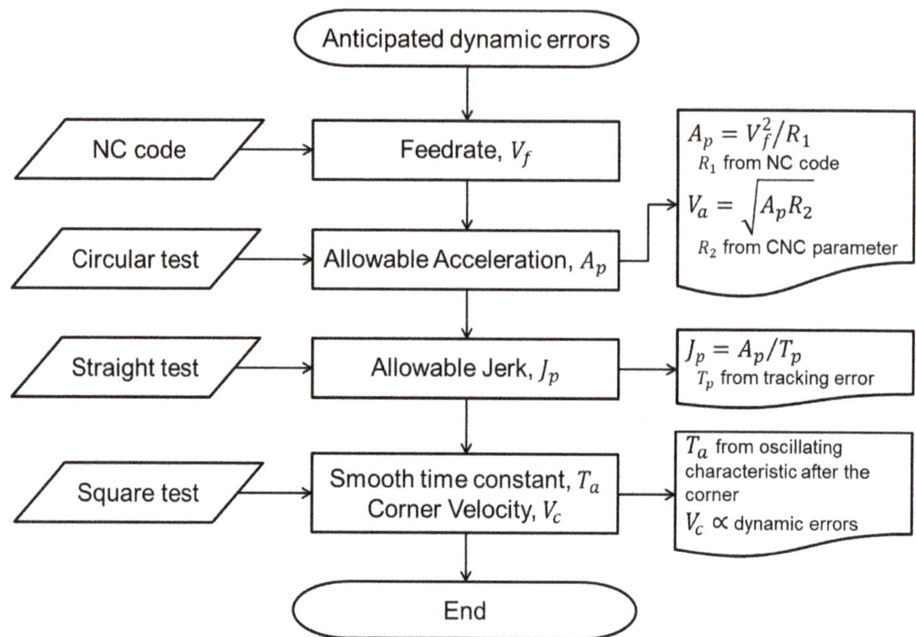

Figure 16. CNC parameter tuning methodology and procedure.

4.1. Allowable Acceleration

The acceleration of any axis results from the force caused by the feed drive system, and leads to contour error on the components. This means that by measuring dynamic errors and correlating them to effective acceleration, maximum allowable acceleration values for each axis can be set depending on the anticipated path accuracy. According to the circular testing, the dynamic errors can be evaluated due to the limited circle radius R. With small circles and increasing V_a, the required acceleration values A_a become very high (Equation (1)). Examples can be seen in Figures 13 and 14. According to Newton's second law and Hooke's law expressed as:

$$F = ma = k\Delta x \tag{5}$$

where m and k are equivalent mass and equivalent stiffness, which are assumed the constants for the closed loop feed drive system in this paper, they are usually hard to tune by the machine operator. The a and Δx are acceleration and deformation of system. According to Equation (5), we can easily understand the relationship between acceleration and dynamic errors. Finally, the equation of acceleration and dynamic errors was obtained by the linear regression method expressed as:

$$A_p = A_a = 63.551 \times \varepsilon_3 - 0.1543 \tag{6}$$

where A_p is allowable acceleration for path movement, A_a is allowable acceleration for axis movement, and ε_3 is anticipated dynamic errors. According to Equation (6), the allowable acceleration has limits of HS/HP/HQ modes: 3.02/0.48/1.12 m/s², respectively. Therefore, these tests are useful for verifying the allowable acceleration value. This means that the CNC parameters related to acceleration can be evaluated. When the NC code was a circular path, the N600 of CNC parameters will be defined by the radius of curvature and Equation (1). In this paper, the radius of curvature of the commercial controller was 5 mm, and the arc velocities in HS/HP/HQ mode were 7373/2939/4490 mm/min, respectively.

4.2. Allowable Jerk

The jerk profiles were similar to the impact force on the mechanical structure. The jerk is a dynamic excitation of the machine, which will cause oscillation in its structure resonance, meaning relative displacements between tool and workpiece, as illustrated in Figure 8. The oscillating characteristic is created by the driving torque or force. This means the measured tracking error of the relevant resonance frequencies can be identified with FFT. For this reason, the allowable jerk values for each axis can be identified at the same time. According to the straight path testing, the mechanical resonance can be identified with tracking error, as illustrated in Figure 9 (blue line). The first resonance of machine was about 38 Hz. The allowable jerk (J_p) can be evaluated due to the time constant of Acc/Dec (T_p) and allowable acceleration (A_p) (maximum value $J_p = A_p/T_p$). The allowable acceleration was set in Section 4.1, and the allowable jerk limits of HS/HP/HQ mode were 114.88/18.29/42.44 m/s^3, respectively. It means that the CNC parameters related to jerk can be evaluated.

4.3. Smoothing Time Constant

The smoothing time constant is similar to the filter, which can smooth the velocity profiles in tool-path planning. The smoothing filter must be designed such that it cancels the mechanical resonance. Its time constant must be set at the inverse of the resonance frequency. According to the square path testing, dynamic errors can be observed at the corner, as illustrated in Figure 17. The oscillation can be seen on the straight line after the corner. In Figure 18, the time constant of the solid line was 0 ms, and the time constant of the dotted line was 26 ms. The result shows that the longer time constant helped to reduce the oscillation of the workpiece surface during machining, but increased contour error and cycle time in machining parts.

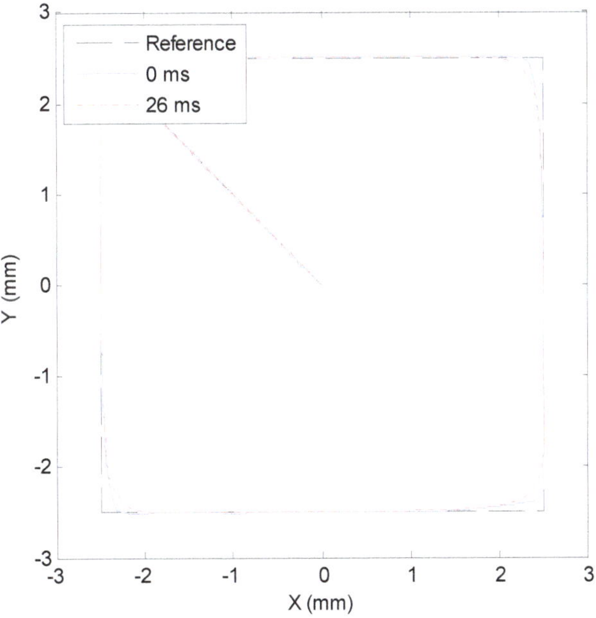

Figure 17. The smoothing time constant led to the contour error.

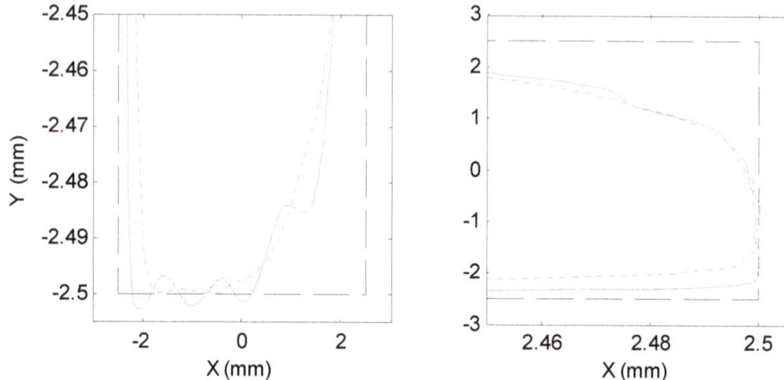

Figure 18. The smoothing time constant led to the oscillation of the tool path.

4.4. Corner Velocity

A higher corner velocity is useful for shorter cycle time, but it would cause larger dynamic errors (or corner error). If it is too high, it easily causes structural vibration at the corners. The vibration could be decreased by the smoothing time constant, but the errors would be increased at the corners. The corner velocity is proportional to the corner error. However, the corner error also includes the effect of the tracking error from the servo loop. However, while there is a lot of domain knowledge to tune the tracking error, it is not easy for the machine operator to use. Therefore, this paper regarded the servo loop as a black box. We proposed the experimental methods to create the relationship between the corner velocity and corner error. According to the experimental results of Section 3.4 for square path testing, the dynamic errors can be observed at the corner. We fed the different jerk, acceleration, V_c, and feedrate for the square path, respectively, summarized as follows:

- The larger jerk and acceleration will increase the corner error as the feedrate increases.
- There are two sets of parameters: One is $J_p = 100$ m/s^3 and $A_p = 2.5$ m/s^2, and the other one is $J_p = 50$ m/s^3, $A_p = 1.2$ m/s^2. The corner errors are more stable than the other set of parameters for the different feedrates.
- If the V_c is set to 3000 mm/min, when the feedrate exceeds 3000 mm/min, the corner errors are more stable. This is caused by the current feedrate being lower than the anticipated V_c. The V_c will give priority to the current feedrate for tool-path planning.

The experimental data on the square path were used to estimate the maximum corner errors of each measurement result. The jerk, acceleration, V_c, and feedrate by the test conditions were considered as inputs, and the corner errors as outputs. Multiple regression analysis was used to obtain the regression formula for the inputs and outputs. Finally, the maximum corner error was set to the dynamic errors of the HS/HP/HQ modes expected in this paper. The feedrate was 4000 mm/min for the square path, the allowable acceleration, and allowable jerk, in Sections 4.1 and 4.2, as the inputs of the regression formula, and solving the allowable V_c of the HS/HP/HQ modes. The corresponding V_c were 1470/250/550 mm/min, respectively.

5. Experimental Validation

In Section 3, the experimental results showed the effectiveness of our proposed machining modes for corresponding CNC parameters, which is relevant to the dynamic errors of the machine. According to the tuning methodology and procedures of the CNC parameters implemented, the relevant CNC parameters were estimated based on the anticipated dynamic errors in Section 4. This section proposed a single square path with the same 4000 mm/min feedrate and the length was 30 mm to verify the validation of the tuning methodology. We used the three sets of CNC parameters (see Table 2) corresponding to the "HS, HP, and HQ modes" in the actual machine platform for validation, thereby observing the performance of the motion dynamic under various modes.

Table 2. CNC parameters for tool-path movement with machining modes.

No.	Sym.	Description	Unit	HS	HP	HQ
N100	J_p	Maximum jerk for tool-path movement	m/s^3	114.88	18.29	42.44
N200	A_p	Maximum acceleration for tool-path movement	m/s^2	3.02	0.48	1.12
N300	T_p	Time constant of Acc/Dec for tool-path movement	ms	26	26	26
N400	T_a	Smoothing time constant for tool-path movement	ms	1	1	26
N500	V_c	Maximum corner velocity for tool-path movement	mm/min	1470	250	550
N600	V_a	Maximum arc velocity for tool-path movement	mm/min	7373	2939	4490

The experimental results are listed in Table 3. We observed the time of the completion of the tool path was the fastest for HS mode. In HP mode, the minimum dynamic errors were of the corners. There was small oscillation that was caused by the tracking error after the cornering in HQ mode. According to the experimental results, the servo parameters could be ignored, and the relevant CNC parameters of the tool-path planning could be adjusted, which can affect the completion time, dynamic errors, and tracking error of the same path.

Table 3. Experimental results for three machining modes.

Description	Unit	HS	HP	HQ
Cycle time	sec	1.935	2.655	2.064
Dynamic errors	mm	0.053	0.018	0.026
Tracking error	mm	0.007	0.003	0.001

The comparison of HS, HP, and HQ modes are listed in Table 4. In the HS mode, the completion time of the path was fastest than the other modes. However, the dynamic errors and the tracking error were increased by 66% and 85%, respectively. In the HP mode, although the completion time of the path was increased by 37% compared with the HS mode and the tracking error was increased by 28%, the HP mode was closed to the ideal path with highest accuracy. In the HQ mode, the surface quality of the workpiece was improved by focusing on reducing the machine resonance, although the dynamic errors were worse than the HP mode. The surface quality improved by 85% compared with the HS mode. However the completion time was only increased by 6% compared with the HS mode.

Table 4. Comparison of three machining modes from the experimental results.

Description	HS	HP	HQ
Cycle time	0%	37%	6%
Dynamic errors	66%	0%	16%
Tracking error	85%	28%	0%

The experimental results verified the tuning methodology for CNC parameters of Section 4. We found that no single set of CNC parameters was able to simultaneously satisfy all of the three machining modes: "High speed", "high accuracy", and "high surface quality". Actually, these three machining modes often contradict each other. For example, we may have the "high productivity", but with "low accuracy and poor surface finish."

6. Conclusions

Since the required product quality affected CNC parameters of the machine tools, the CNC parameters must be selected to vary the machining performance of the tool-path planning, named as

HS, HP, and HQ modes, respectively. Actually, the manufacturer provided only a single set of CNC parameters for the commercial CNC machine tools. It was not enough for the different machining requirements. However, the performance of the servo loop and actual machine was fixed and was hard to be tuned, another reason for needing to have quite a domain knowledge to tune the CNC parameters' requirements. Therefore, the operator of the machine cannot easily process the different machining requirements. This paper introduced the relationship of CNC parameters, dynamic errors, and machining modes. The dynamic errors defined the difference of the ideal position for NC code to the actual position for linear scale in this study. According to the anticipated dynamic errors for the three machining modes and feed motion testing by several specified paths' experiment, such as straight, square, and circular paths in actual machine tools, the HS, HP, and HQ modes can be achieved by properly tuning the CNC parameters via the tuning methodology. The experimental results proposed an approach that contained three sets of CNC parameters. The CNC parameters were selected to be larger values of J_p, A_p, V_c, and V_a, the path completion time was shortened, and the processing efficiency improved to match the HS mode. However, the corresponding dynamic errors became larger, which sacrificed accuracy and surface quality. On the other hand, the smaller values of CNC parameters resulted in larger completion time and smaller tracking error, and the geometric accuracy was improved to reach the HP mode. The HQ mode dealt with the worse surface quality of the product, caused by feed force for the feed motion. When the feed motion experienced oscillation on the corner or circular paths, the exact T_a reduced the impact of feed force during the high Acc/Dec period. Although the dynamic errors of HQ mode were larger than in the HP mode, the oscillation could be reduced during the feed motion. If there is larger oscillation occurring in the HS mode, in addition to the reduction in J_p, A_p, V_c, and V_a, it can also use the T_a to decrease the oscillation for the smooth surface quality of the geometric shape. However, the cycle time became longer, caused by T_a in the HQ mode. The experimental results were introduced to illustrate the effectiveness of the proposed approach. The advantage of this paper is that it has developed a tuning methodology and procedure of the CNC parameters based on their relevant machining mode required; it can easily be implemented in a commercial CNC controller of the machine tools. The machine user can simply select the best machining mode according to the machining requirements.

Author Contributions: B.-F.Y. and J.-S.C. In itiated and developed the concepts related to this research work. B.-F.Y. performed the collected data for the designed experiment and analyzed the data. Both of them discussed the experimental results and developed the tuning methodology. B.-F.Y. wrote the paper draft under Chen's guidance. All authors discussed the results and commented on the manuscript. All authors have read and agreed to the published version of the manuscript.

Funding: The authors would like to thank the reviewers for their suggestions. This research was supported by Ministry of Science and Technology, Taiwan grant to MOST-108-2218-E-005-021-"Intelligent multi-axis machine tool technology for machining the advanced materials of the semiconductor and electronics industries".

Conflicts of Interest: The authors declare no conflict of interest.

References

1. Erkorkmaz, K. High Speed Contouring Control for Machine Tool Drives. Master's Thesis, The University of British Columbia, Vancouver, BC, Canada, February 1999.
2. Erkorkmaz, K.; Altintas, Y. High speed CNC system design. Part I: Jerk limited trajectory generation and quantic spline interpolation. *Int. J. Mach. Tools Manuf.* **2001**, *41*, 1323–1345. [CrossRef]
3. Nam, S.-H.; Yang, M.-Y. A study on a generalized parametric interpolator with real-time jack-limited acceleration. *Comput. Aided Des.* **2004**, *36*, 27–36. [CrossRef]
4. Barre, P.-J.; Bearee, R.; Borne, P.; Dumetz, E. In fluence of a Jerk Controlled Movement Law on Vibratory Behaviour of High-Dynamics Systems. *J. In tell. Robot. Syst.* **2005**, *42*, 275–293. [CrossRef]
5. Lin, M.-T.; Tsai, M.-S.; Yau, H.-T. Development of a dynamics-based NURBS interpolator with real-time look-ahead algorithm. *Int. J. Mach. Tools Manuf.* **2007**, *47*, 2246–2262. [CrossRef]
6. Yau, H.-T.; Wang, J.-B. Fast Bazier interpolator with real-time lookahead function for high-accuracy machining. *Int. J. Mach. Tools Manuf.* **2007**, *47*, 1518–1529. [CrossRef]

7. Sencer, B.; Altintas, Y.; Croft, E. Feed optimization for five-axis CNC machine tools with drive constraints. *Int. J. Mach. Tools Manuf.* **2008**, *48*, 733–745. [CrossRef]

8. Tapie, L.; Mawussi, K.B.; Anselmetti, B. Circular tests for HSM machine tools: Bore machining application. *Int. J. Mach. Tools Manuf.* **2007**, *47*, 805–819. [CrossRef]

9. Pateloup, V.; Duc, E.; Ray, P. Corner optimization for pocket machining. *Int. J. Mach. Tools Manuf.* **2004**, *44*, 1343–1353. [CrossRef]

10. Andolfatto, L.; Lavernhe, S.; Mayer, J.R.R. Evaluation of servo, geometric and dynamic error sources on five-axis high-speed machine tool. *Int. J. Mach. Tools Manuf.* **2011**, *51*, 787–796. [CrossRef]

11. Lee, K.; Ibaraki, S.; Matsubara, A.; Kakino, Y.; Suzuki, Y.; Arai, S.; Braasch, J. A Servo Parameter Tuning Method for High-Speed NC Machine Tools based on Contouring Error Measurement. In *Laser Metrology and Machine Performance VI*; WIT Press: Southampton, UK, 2002.

12. Parenti, P.; Bianchi, G.; Cau, N.; Albertelli, P.; Monno, M. A Mechatronic Study on a Model-Based Compensation of Inertial Vibration in a High-Speed Machine Tool. *J. Mach. Eng.* **2011**, *11*, 91–104.

13. Bringmann, B.; Maglie, P. A method for direct evaluation of the dynamic 3D path accuracy of NC machine tools. *CIRP Ann.* **2009**, *58*, 343–346. [CrossRef]

14. Li, B.; Luo, B.; Mao, X.; Cai, H.; Peng, F.; Liu, H. A new approach to identifying the dynamic behaviour of CNC machine tools with respect to different worktable feed speeds. *Int. J. Mach. Tools Manuf.* **2013**, *72*, 73–84. [CrossRef]

15. Franco, P.; Estrems, M.; Faura, F. In fluence of radial and axial runouts on surface roughness in face milling with round insert cutting tools. *Int. J. Mach. Tools Manuf.* **2004**, *44*, 1555–1565. [CrossRef]

16. Buj-Corral, P.; Vivancos-Calvet, V.; González-Rojas, H. Roughness variation caused by grinding errors of cutting edges in side milling. *Mach. Sci. Technol.* **2013**, *17*, 575–592. [CrossRef]

17. Vivancos-Calvet, S.; Wiackiewicz, M.; Krolczyk, G.M. Study on metrological relations between instant tool displacements and surface roughness during precise ball end milling. *Measurement* **2018**, *129*, 686–694.

18. Mia, M.; Królczyk, G.; Maruda, R.; Wojciechowski, S. In telligent Optimization of Hard-Turning Parameters Using Evolutionary Algorithms for Smart Manufacturing. *Materials* **2019**, *12*, 879. [CrossRef]

19. Li, K.-Y.; Luo, W.-J.; Yang, M.-H.; Hong, X.-H.; Luo, S.-J.; Chen, C.-H. Effect of Supply Cooling Oil Temperature in Structural Cooling Channels on the Positioning Accuracy of Machine Tools. *J. Mech.* **2019**, *35*, 887–900. [CrossRef]

20. Li, K.-Y.; Luo, W.-J.; Huang, J.-Z.; Chan, Y.-C.; Pratikto; Faridah, D. Operational Temperature Effect on Positioning Accuracy of a Single-Axial Moving Carrier. *Appl. Sci.* **2017**, *7*, 420. [CrossRef]

21. Altintas, Y.; Verl, A.; Brecher, C.; Uriarte, L.; Pritschow, G. Machine tool feed drives. *CIRP Ann.* **2011**, *60*, 779–796. [CrossRef]

22. Altintas, Y. Virtual Machine Tool. *CIRP Ann.* **2005**, *54*, 115–138. [CrossRef]

23. Erkorkmaz, K.; Altintas, Y.; Yeung, C.-H. Virtual computer numerical control system. *CIRP Ann.* **2006**, *55*, 399–402. [CrossRef]

24. Lin, C.-Y.; Lee, C.-H. Remote Servo Tuning System for Multi-Axis CNC Machine Tools Using a Virtual Machine Tool Approach. *Appl. Sci.* **2017**, *7*, 776. [CrossRef]

25. Christian, B.; Fey, M.; Baumler, S. Damping models for machine tool components of linear axes. *CIRP Ann. Manuf. Technol.* **2013**, *62*, 399–402.

26. Zaeh, M.F.; Baudisch, T. Simulation environment for designing the dynamic motion behaviour of the mechatronic system machine tool. *Proc. In st. Mech. Eng. Part B J. Eng. Manuf.* **2003**, *217*, 1031–1035. [CrossRef]

27. Heidenhain. *Technical Manual iTNC530 HSCI*; Heidenhain: Traunreut, Germany, 2011.

28. Siemens. SINUMERIK 840D sl Basic Functions–Function Manual. Siemens AG, Germany. 2013. Available online: https://support.industry.siemens.com/cs/document/109763231/sinumerik-840d-sl-basic-functions?dti=0&lc=en-WW (accessed on 13 April 2020).

Article

An Accurate and Efficient Approach to Calculating the Wheel Location and Orientation for CNC Flute-Grinding

Yang Fang [1,2,3], Liming Wang [1,2,3,*], Jianping Yang [1,2,3] and Jianfeng Li [1,2,3]

1   School of Mechanical Engineering, Shandong University, 17923 Jingshi Road, Jinan 250061, China; fangyang@mail.sdu.edu.cn (Y.F.); yangjianping_yjp@mail.sdu.edu.cn (J.Y.); ljf@sdu.edu.cn (J.L.)
2   National Demonstration Center for Experimental Mechanical Engineering Education, Shandong University, 17923 Jingshi Road, Jinan 250061, China
3   Key Laboratory of High Efficiency and Clean Mechanical Manufacture, Shandong University, Ministry of Education, Jinan 250061, China
*   Correspondence: liming_wang@sdu.edu.cn; Tel.: +86-531-88392208

Received: 27 May 2020; Accepted: 16 June 2020; Published: 19 June 2020

Abstract: The profile of flutes has a great influence on the stiffness and chip-removal capacity of end-mills. Generally, the accuracy of flute parameters is determined by the computer numerical control (CNC) grinding machine through setting the wheel's location and orientation. In this work, a novel algorithm was proposed to optimize the wheel's location and orientation for the flute-grinding to achieve higher accuracy and efficiency. Based on the geometrical constraint that the grinding wheel should always intersect with the bar-stock while grinding the flutes, the grinding wheel and bar-stock were simplified as an ellipse and circle via projecting in the cross-section. In light of this, we re-formulated the wheel's determination model and analyzed the geometrical constraints for interference, over-cut and undercut in a unified framework. Then, the projection model and geometrical constraints were integrated with the evolution algorithm (i.e., particle swarm optimization (PSO), genetic algorithm (GA) for the population initialization and local search operator so as to optimize the wheel's location and orientation. Numerical examples were given to confirm the validity and efficiency of the proposed approach. Compared with the existing approaches, the present approach improves the flute-grinding accuracy and robustness with a wide range of applications for various flute sizes. The proposed algorithm could be used to facilitate the general flute-grinding operations. In the future, this method could be extended to more complex grinding operations with the requirement of high accuracy, such as various-section cutting-edge resharpening.

Keywords: flute-grinding; evolution algorithms; wheel location and orientation

---

1. Introduction

Flutes, as the major structure of end-mills, play an important role in the cutting performance [1–4]. A flute can be defined by the following three parameters: core radius, flute angle and rake angle [5–7]. The rake angle influences the cutting force, while the core radius and the helix angle determine the stiffness and chip-removal capacity of the cutters. The performance of those flute parameters quietly depends on the manufacturing accuracy. Generally, the flute is manufactured by the CNC grinding machine through the determination of location and orientation for the grinding wheel path [8–10]. In the flute-grinding operations, the grinding wheel will move with a helix path to generate the grooves. In recent years, much attention has been paid to developing an advanced wheel path determination model and optimized algorithms for the CNC flute grinding to minimize the manufacture errors and improve calculation efficiency.

The kinematics of flute-grinding operations for CNC grinders has been developed by many researchers. For instance, Kim et al. [11] developed a simulation method with Boolean operations to construct the helix motion and developed an iterative process to compute the wheel geometry and location data. Although this method could be used for virtual cutting tests in the CAM system, it is time-consuming to achieve a high machining accuracy. To improve the precision of generated flutes, Li [12] established a novel algorithm to calculate the numerical data of flutes based on the enveloping theory, in which the flute profile was interpolated with appropriate discrete points using the cubic polynomial expression. However, this method might be invalid for grinding with a bevel-type wheel. In order to calculate the location and orientation of wheels with complex shapes (e.g., 1B1, 1E1, 1F1, and 4Y1 wheels), Habibi et al. [13] used virtual grinding curves to formulate the grinding processes. Based on the virtual curves, they calculated the grinding error with the worn wheels and compensated the wheel path. Generally, the kinematic of flute-grinding was represented by several transcendental equations, which were supposed to be solved to get the generated helical flute. However, it is very complicated to give the analytical solution, and thus numerical analysis was generally used to describe the flute profile in current studies, which suffer from long computation time. For free-form grinding wheels, Wasif et al. [14] presented a novel method for five-axis CNC grinding through the optimization of the grinding wheel geometry, which was constructed with line segments and circular arcs. Although this method could economically produce or dress the grinding wheel for accurately grinding the end-mill cutters, it ignored the optimization of the wheels' path. On the basis of the flute-grinding model, the following problem is how to achieve the desired flute parameters by setting the trajectory of the grinding wheel. In industry, the conventional way is to grind end-mill flutes by trial and error, which is costly and time-consuming [15]. Mathematically, the desired flute profile can be viewed as an optimization problem with regards to the wheel's shape and configuration. To solve this problem, Chen et al. [16] proposed an iteration algorithm to determine the wheel location and orientation. For each loop, the generated flute parameters were compared with the desired values until they converged within the target range. The iteration method has a high calculating speed and precision, but it required a proper initial value, which cannot be easily determined without experience. In Karpuschewski's research [17], particle swarm optimization (PSO) was investigated to search the wheel location for a given helical flute and grinding wheel profile. Recently, Li et al. [18] extended this work with a novel graphic analysis method and niche particle swarm optimization (NPSO) algorithm to solve the problem for multi-objective of machining accuracy. Although the evolution algorithms (EA, i.e., PSO, GA) could be used in the wheel path optimization process with a global search strategy for such a complex nonlinear problem, they are not stable in convergence, especially for some small flutes (flute diameter < 1 mm).

In view of the above survey, it can be seen that the current study has addressed the modeling of flute-grinding well, but fast and stable algorithms are still required for further study. Currently, extreme-size cutters are widely used in industry, such as the micro-milling cutter or turbine blade root milling cutter. The flute parameters for those cutters will greatly affect their cutting performance, which requires higher accuracy. However, the current algorithms lack the definition of various constraint conditions in machining, which cannot guarantee the machining accuracy and calculation stability. In this work, a new method is presented to calculate the flute parameters and determine its CNC grinding operations. Compared to the above studies, we projected the grinding wheel in the cross-section and used a two-parameter operator to control the grinding operations. Regarding the wheel path optimization algorithm, the projection model was integrated with the evolution algorithm (i.e., PSO, GA) for the population initialization and local search operator. Compared to the current EA method, the improved method showed better stability and short computation time. In addition, it could achieve higher accuracy for a wide range of flute sizes with various profiles, especially for the above-mentioned small flutes.

The outline of the paper is described as follows. In Section 2, the kinematic of flute-grinding is a model with an explicit expression, and the generated flute parameters are formulated. In Section 3, a projection model is proposed for the grinding processes and a two-parameter operator to control the wheel location and orientation is introduced. Additionally, the constraints for undercut, over-cut and extreme-cut of flutes are investigated. This projection model will integrate with the subsequent evolution algorithm. Section 4 presents a wide range of flute-grinding problems to test the accuracy, efficiency, and robustness of the proposed method. Finally, Section 5 summarizes the whole article and points out the contribution of our work.

2. Modeling of Flute-grinding Processes

2.1. Grinding Wheel Modeling

In this work, a standard conical grinding wheel was applied to implement the flute-grinding. Figure 1 illustrates this type of standard grinding wheel. To describe the wheel's geometry, a frame O_G, denoted as a wheel coordinate system, is developed and shown in Figure 1. Then, the parametric representation of the wheel $^G W(h, \theta)$ in the wheel coordinate system can be derived as

$$^G W(h, \theta) = \begin{bmatrix} f \cdot \cos \theta \\ f \cdot \sin \theta \\ h \end{bmatrix}, \tag{1}$$

where $h \in [0, H]$, $\theta \in [0, 2\pi]$ and $f = R - h \cot \alpha$.

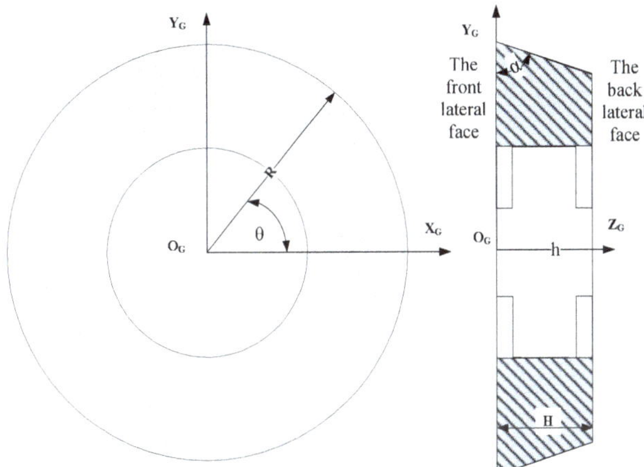

Figure 1. Modeling of conical grinding wheel.

Except for the wheel surface, the lateral face will also be involved in the grinding processes. Geometrically, the front lateral face $^G W(0, \theta)$ and the back lateral face $^G W(H, \theta)$ can be obtained by setting $h = 0$ and $h = H$. Additionally, the wheel surface normal is deduced from Equation (1) as follows:

$$^G N = \begin{bmatrix} \cos \theta \\ \sin \theta \\ \cot \alpha \end{bmatrix} \tag{2}$$

2.2. Kinematic of CNC Flute-Grinding

To demonstrate the flute-grinding operations, another framework O_T is established in Figure 2, which is denoted as a tool coordinate system. In the modeling, O_T is static while O_G is moving with the grinding wheel. The flute-grinding operations consist of two steps: (1) wheel set-up and (2) wheel moving with a helix trajectory.

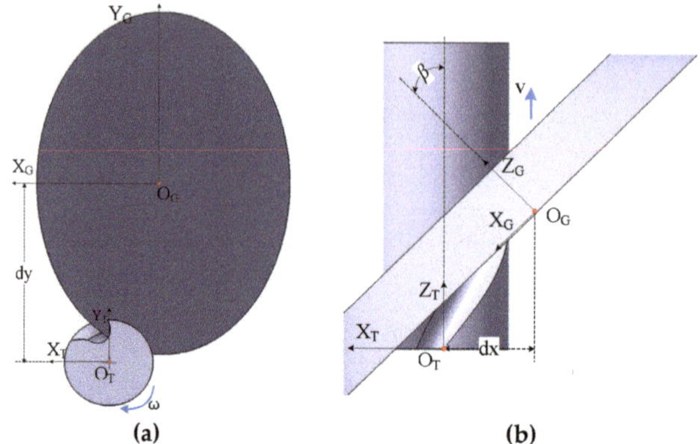

Figure 2. Flute-grinding processes: (**a**) $X_T Y_T$ view; (**b**) $X_G Y_G$ view.

In the wheel set-up operation, the wheel is configured in $\mathbf{O_T}$ with a specified location and orientation shown in Figure 2. The wheel location is defined by the wheel center $\mathbf{O_G}$, which is denoted by the coordinate value $[dx \; dy \; dz]$. The wheel orientation is defined as an angle, which can be viewed as rotating the grinding wheel about the $\mathbf{Y_T}$ axis in a counter clockwise direction through the angle β. To this end, the set-up operation in the tool coordinate system can be expressed using the homogeneous transformation matrix in Equation (3).

$$
{}^T\mathbf{M}_1 = \begin{bmatrix} \cos\beta & 0 & \sin\beta & dx \\ 0 & 1 & 0 & dy \\ -\sin\beta & 0 & \cos\beta & dz \\ 0 & 0 & 0 & 1 \end{bmatrix}.
\tag{3}
$$

In the wheel helix motion, the wheel moves along the $\mathbf{Z_T}$ axis with a translation velocity v while the cutter rotates about the $\mathbf{Z_T}$ axis with a velocity ω. In the tool coordinate system, the kinematics matrix of the helix motion is represented as follows:

$$
{}^T\mathbf{M}_1 = \begin{bmatrix} \cos(\omega \cdot t) & 0 & -\sin(\omega \cdot t) & 0 \\ \sin(\omega \cdot t) & 1 & \cos(\omega \cdot t) & 0 \\ 0 & 0 & 1 & v \cdot t \\ 0 & 0 & 0 & 1 \end{bmatrix},
\tag{4}
$$

where t represents the grinding time.

To guarantee the helix angle λ, the translation v and the rotation ω is supposed to satisfy the following condition:

$$
\cot\lambda = \frac{v}{r_T \cdot \omega}.
\tag{5}
$$

Based on the above operations, the kinematic of grinding wheels can be obtained with regarding machining time t in the tool coordinate system by integrating Equations (1)–(5), listed in Equation (6). In addition, to simplify the calculation, the rotation speed ω is generally set as 1 in the following equations.

$$
\begin{aligned}
{}^TW(h,\theta,t) &= {}^TM_2 \cdot {}^TM_1 \cdot {}^GW(h,\theta) = \\
&\begin{bmatrix} dx \cdot \cos t - dy \cdot \sin t + h \cdot \sin\beta \cdot \cos t - f \cdot \sin\theta \cdot \sin t + f \cdot \cos\beta \cdot \cos\theta \cdot \cos t \\ dy \cdot \cos t + dx \cdot \sin t + h \cdot \sin\beta \cdot \sin t + f \cdot \sin\theta \cdot \cos t + f \cdot \cos\beta \cdot \cos\theta \cdot \sin t \\ h \cdot \cos\beta + v \cdot t - f \cdot \sin\beta \cdot \cos\theta + dz \end{bmatrix}.
\end{aligned}
\tag{6}
$$

Geometrically, the flutes are generated by the envelope of grinding wheels. The envelope surface consists of a group of curves, which are called the contact curve [19]. As shown in Figure 3, the contact curve is composed of two parts. One is generated by the wheel surface, which can be deduced using envelope theory. The other is formed by part of the wheel edges.

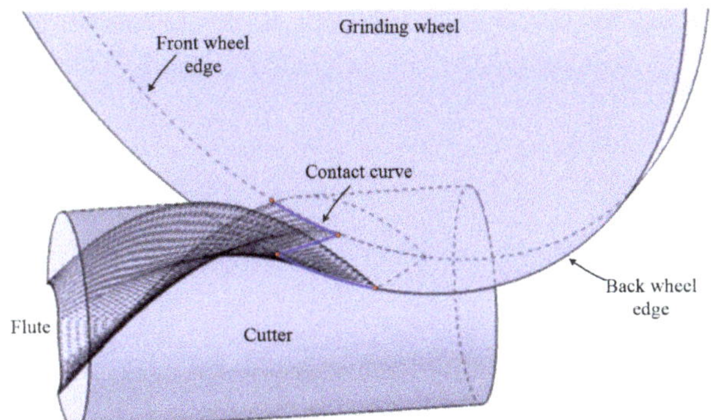

Figure 3. Illustration of the flute generation and the contact curve.

For the first part of the contact curve, it can be obtained using the conjugate theory as shown in Equation (7):

$$^{T}N \cdot {}^{T}V = 0, \tag{7}$$

where $^{T}N = {}^{T}M_{2} \cdot {}^{T}M_{1} \cdot {}^{G}N$ and $^{T}V(h,\theta,t) = \dfrac{d\left({}^{T}W(h,\theta,t)\right)}{dt}$.

By solving Equation (7), the equation of this contact curve can be deduced as Equation (8).

$$
\begin{aligned}
v \cdot (dz - \sin\beta \cdot \cos\theta + \cot\alpha \cdot \cos\beta) - (dy + \sin\theta \cdot f) \cdot (dx + \cot\alpha \cdot \sin\beta + \cos\beta \cdot \cos\theta) \\
+ (dy + \sin\theta) \cdot (dx + h \cdot \sin\beta + \cos\beta \cdot \cos\theta \cdot f) = 0.
\end{aligned}
\tag{8}
$$

By solving this triangular equation, the explicit expression of the contact curve can be obtained in Equation (9):

$$\theta^{*} = atan2(B, A) + atan2\left(\sqrt{A^{2} + B^{2} - C^{2}}, C\right), \tag{9}$$

where
$$
\begin{cases}
A = -dy \cdot \cos\beta - v \cdot \sin\beta \\
B = dx + f \cdot \sin\beta \cdot \cot\alpha \\
C = dy \cdot \cot\alpha \cdot \sin\beta - dz \cdot v - v \cdot \cot\alpha \cdot \cos\beta
\end{cases}
.
$$

Substituting Equation (9) into Equation (6), the first part of the flute surface can be obtained, which is formed by the envelope of the wheel surface in a general form in Equation (10).

$$
^{T}W(h,t) =
\begin{bmatrix}
dx \cdot \cos t - dy \cdot \sin t + h \cdot \sin\beta \cdot \cos t - f \cdot \sin\theta^{*} \cdot \sin t + f \cdot \cos\beta \cdot \cos\theta^{*} \cdot \cos t \\
dy \cdot \cos t + dx \cdot \sin t + h \cdot \sin\beta \cdot \sin t + f \cdot \sin\theta^{*} \cdot \cos t + f \cdot \cos\beta \cdot \cos\theta^{*} \cdot \sin t \\
h \cdot \cos\beta + v \cdot t - f \cdot \sin\beta \cdot \cos\theta^{*} + dz
\end{bmatrix}
\tag{10}
$$

The other part of the contact curve generated by wheel edge can be obtained by setting $h = 0$ or $h = H$ for Equation (6) denoted as $^{T}W(0,\theta,t)$ and $^{T}W(H,\theta,t)$.

Generally, the flute parameters are presented in the cross-section with the definition of core radius, flute angle and rake angle [20,21]. The flute profile could be easily obtained by setting $Z = 0$ for Equation (10), as shown in Equation (11):

$$^T W(h) = \begin{bmatrix} x_c \\ y_c \end{bmatrix} =$$

$$\begin{bmatrix} dx \cdot \cos t^* - dy \cdot \sin t^* + h \cdot \sin\beta \cdot \cos t^* - f \cdot \sin\beta \cdot \sin t^* + f \cdot \cos\beta \cdot \cos\theta^* \cdot \cos t^* \\ dy \cdot \cos t^* + dx \cdot \sin t^* + h \cdot \sin\beta \cdot \sin t^* + f \cdot \sin\beta \cdot \cos t^* + f \cdot \cos\beta \cdot \cos\theta^* \cdot \sin t^* \end{bmatrix} , \tag{11}$$

where $t^* = \frac{R \cdot \sin\beta \cdot \cos\theta^* - h \cdot \cos\beta - dz}{v}$.

As shown in Figure 4a, the flute parameters are defined as follows:

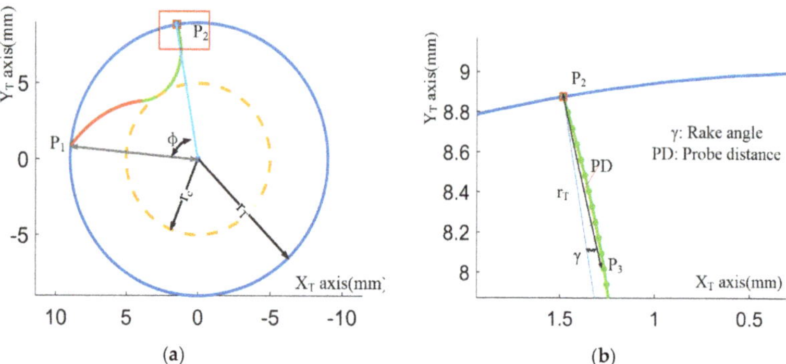

Figure 4. The flute parameters: (**a**) flute angle and core radius; (**b**) rake angle.

The flute angle is established as the opening angle between the vector $\mathbf{O_T P_1}$ and another vector $\mathbf{O_T P_2}$, which can be calculated using: $\phi = \mathrm{acos}(\mathbf{O_T P_1} \cdot \mathbf{O_T P_2})$.

The core radius is the minimum distance from a point $\mathbf{O_T}$ to the flute profile, which can be calculated as: $r_c = \min\left(x_c^2 + y_c^2\right)$.

The rake angle is illustrated in Figure 4b, which is a close-up of Figure 4a at point $\mathbf{P_1}$. In practice, the rake angle is measured at the start point $\mathbf{P_3}$ with a measure distance **PD** (**PD** is set as 5% of tool radius in this work). Geometrically, the rake angle is the included angle of the two vectors $\mathbf{P_2 O_T}$ and $\mathbf{P_2 P_3}$, which can be expressed as: $\gamma = \mathrm{acos}(\mathbf{P_2 O_T} \cdot \mathbf{P_2 P_3})$

In addition, the points $\mathbf{P_1}$, $\mathbf{P_2}$ and $\mathbf{P_3}$ can be obtained by the following conditions:

(1) Point $\mathbf{P_1}$ is deduced by Equation (10) satisfying the condition $\sqrt{x_c^2 + y_c^2} = r_T$;

(2) Point $\mathbf{P_2}$ is deduced by setting $h = 0$ to Equation (10), and satisfying the condition $\sqrt{x_c^2 + y_c^2} = r_T$;

(3) Point $\mathbf{P_3}$ is deduced by setting $|\mathbf{P_2 P_3}| = \mathrm{PD} = 5\% \cdot r_T$.

3. Determination of Wheel Location and Orientations with a 2D Projection

Mathematically, the flute-grinding operations can be simplified with three equations shown in Equation (12). For these equations, the wheel location $[dx\ dy]$ and orientation β are supposed to be calculated to configure the wheel path and generate the designed flute profile. Generally, the intelligent evolution algorithms, i.e., GA or PSO, were used to solve the above equations. However, it was reported that the selection of initial points would greatly affect the accuracy and efficiency of the optimization. In practice, we found that the initial wheel's location and orientation were confined by several geometrical conditions, such as inference avoidance, contact constraints, etc., which can be used to define the feasible space for the initial points. To this end, an optimization method was introduced to build the feasible space and constraints by mapping the grinding operations into a two-dimensional projection.

$$\begin{cases} f_{rake}\,(dx, dy, \beta) = \gamma_0 \\ f_{flute}\,(dx, dy, \beta) = \phi_0 \\ f_{core}\,(dx, dy, \beta) = r_{c0} \end{cases} . \tag{12}$$

3.1. Grinding Operation Projection

Geometrically, the flutes are formed by the intersection between the wheel and the bar-stock. To represent the intersection, the grinding wheel and the bar-stock were projected in the cross-section, which is shown in Figure 5. The bar-stock is simplified as a circle area and the wheel edge can be expressed with an ellipse area. The ellipse could be represented with regard to the wheel's location and orientation denoted as $r_e|_{(dx,dy,\beta)}$. In order to assure the accuracy of the core radius, the ellipse will intersect with the circle r_T and tangent with the circle r_c at a tangent point. The tangent point is defined by the parameter θ_c, denoted by $P_c|_{(\theta_c)}$. It can be seen that the point $P_c|_{(\theta_c)}$ can be used to locate the ellipse. The algebra relation between $P_c|_{(\theta_c)}$ and $r_e|_{(dx,dy,\beta)}$ is deduced in Appendix A. In light of the above geometrical relation, the wheel path determination equation can be re-organized concerning the parameters θ_c and β, shown in Equation (13).

$$\begin{cases} f_{rake}\,(\theta_C, \beta) = \gamma_0 \\ f_{flute}\,(\theta_C, \beta) = \phi_0 \\ f_{core}\,(\theta_C, \beta) = r_{c0} \end{cases} , \tag{13}$$

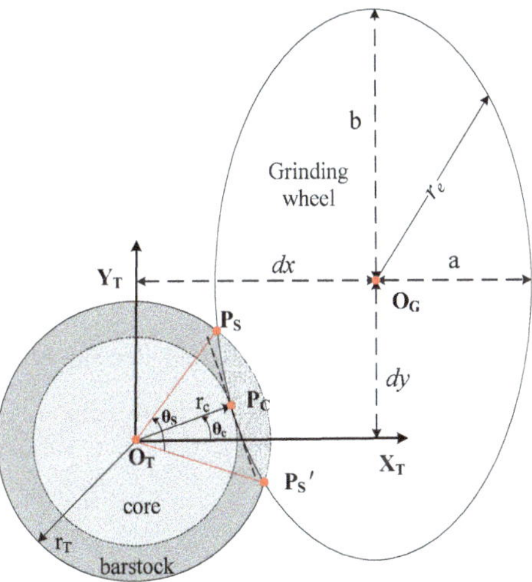

Figure 5. Illustration of the projection of the wheel edge and cutter profile within the cross-section.

To solve this equation, an optimization model is defined as following:

$$min\xi(\theta_C, \beta) = s.t.\xi = max\left\{ \left|\frac{f_{rake}(\theta_C, \beta) - \gamma_0}{\gamma_0}\right|, \left|\frac{f_{flute}(\theta_C, \beta) - \phi_0}{\phi_0}\right|, \left|\frac{f_{core}(\theta_C, \beta) - r_{c0}}{r_{c0}}\right| \right\}, \tag{14}$$

where ξ is the grinding error in the following description.

In addition, several constraints were defined as follows:

- **Constraint 1:** the tangent point $P_c|_{(\theta_c)}$ should always locate in the first or the second quadrant,

$$\theta_C \in [0, \pi]. \tag{15}$$

- **Constraint 2:** the wheel edge cannot be separated with the bar-stock and overcut the core radius,

$$r_c(\theta_C, \beta) = |O_T P_C| \in [r_{c0}, r_T], \tag{16}$$

- **Constraint 3:** to avoid interference, the open-angle θ_S should satisfy the following condition (see Appendix B),

$$\theta_S \geq \Omega, \tag{17}$$

To sum up, the flute-grinding operations can be re-formulated by the following optimization problem:

$$\min \xi(\theta_C, \beta),$$

subjected to: Equations (15)–(17).

Compared with current flute-grinding optimization models, this project method has two advantages: (1) instead of three decision parameters (dx, dy, β), only two parameters (θ_C, β) were required to be considered, which will simplify the calculation during the iterations; (2) three constraints could be used to confine the feasible area and generate the proper initial points, which would improve the robustness and efficiency of the optimization.

3.2. Calculation Procedure with the Improved GA and PSO

To solve the above-constrained optimization problem, the projection model with the GA and PSO was integrated to calculate the wheel's location and orientation for flute-grinding operation. As mentioned, the initial points would greatly affect the optimization results. In this work, an initial points generation algorithm was proposed in Algorithm 1, which could be used for the population initialization for GA or PSO. It is noted that all the initial points would be checked by the over-cut and interference constraints in the projection model. Furthermore, the flowchart of improved GA and PSO (IGA, IPSO) integrated with the projection model is shown in Figure 6. First, a set of initial points were generated with the algorithm in, which is used to calculate the following wheel's location and the generated flute parameters. In light of this, the grinding errors were evaluated and set as the fitness function for the GA and PSO. For the IGA and IPSO, the off springs or particles generated by the iteration operators, e.g., mutation or crossover, will be checked by the proposed over-cut and interference constraints. The iteration will stop while satisfying either of the following conditions: (1) iterations n > 100 (2) there is no obvious improvement in the grinding errors within n = 10 succeeding iterations. With this procedure, the optimized wheel's location and orientation, the generated flute parameters and grinding errors could be obtained.

Algorithm 1 Generate N initial points

Input: Desired flute $\{r_T, \lambda, r_{c0}, \gamma_0, \phi_0\}$ and wheel parameters $\{R, H, \alpha\}$
Output: initial points (pop) for β & θ_C
1.　　$pop = \{\varphi\}$.
2. **while** n < N **do**
3.　　　　$\beta = \mathrm{random}(0, \pi/2)$ and $\theta_C = \mathrm{random}(0, \pi)$.
4.　　　　Calculate $r_c(\theta_C, \beta)$ and θ_s in the projection model (see Appendix A)
5.　　　　**If** satisfy the constraints for Equation (16) and Equation (17).
6.　　　　$pop = pop \cup \{\beta, \theta_C\}$
7.　　　　else
8.　　　　　　go to line 3
9.　　　　**end if**
10. **return** pop

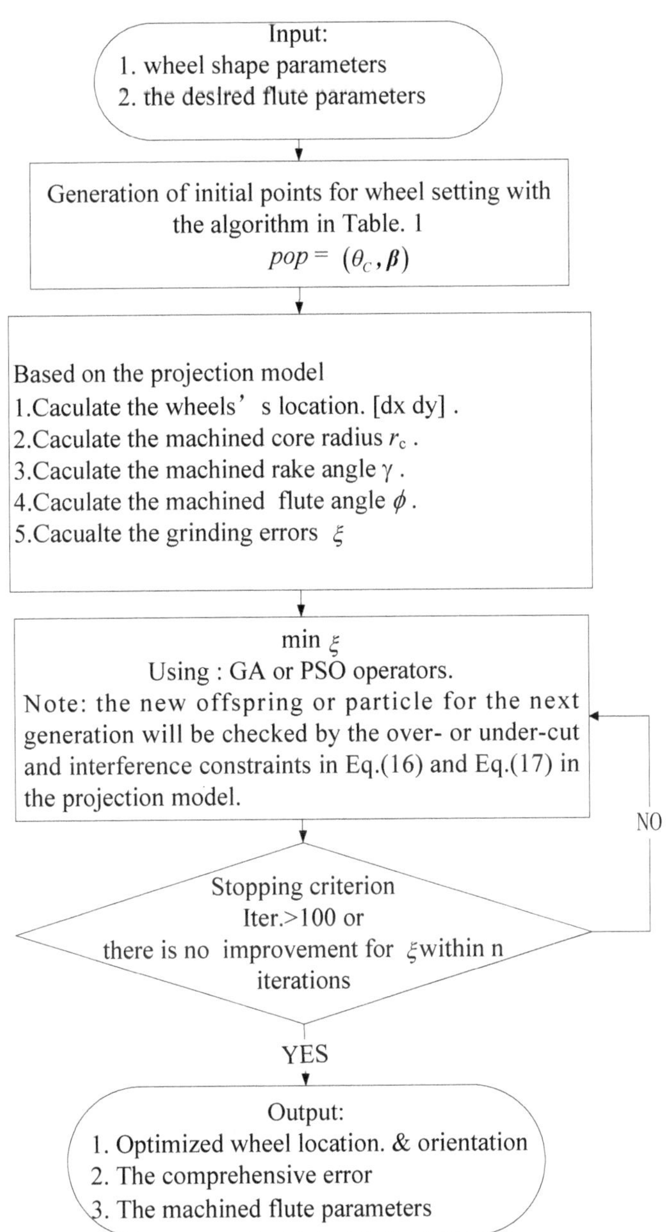

Figure 6. Flow chart of the improved genetic algorithm (IGA) and improved particle swarm optimization (IPSO) with the projection model.

4. Numerical Simulation

To test the accuracy and efficiency of the presented model, the numerical simulation was conducted with various sizes of flutes. Three types of grinding wheels were provided in Table 1. The designed flutes were divided into three groups according to their size: small ($r_T \leq 1$ mm), medium (1 mm $\leq r_T \leq 20$ mm), large ($r_T \geq 20$ mm). To grind those flutes, the wheel's location and orientation in CNC operations were determined with the proposed improved GA and PSO (IGA, IPSO), and also compared with the traditional GA and PSO. For each instance, the optimization was run 10 times and the average results and deviation were recorded. The optimization program was implemented in MATLAB 2010 on a computer with Intel Core i5, 2.39 GHz, 4 GB RAM. The parameters of the GA and PSO were set in Table 2. The desired flutes parameters are listed in Table 3. The helix angle for those flutes was set as 30 degrees.

Table 1. Specification of grinding wheel.

Wheel Parameters	Wheel 1	Wheel 2	Wheel 3
wheel width H (mm)	5	20	40
wheel radius R (mm)	30	75	75
wheel angle α (deg.)	75	75	90

Table 2. Parameters of optimization algorithms.

Algorithm	Parameters Setting
GA or IGA	initial population size: 100 range of crossover probability: 0.2 range of mutation probability: 0.1 stopping condition: iterations > 100 or $\xi < 1 \times 10^{-4}$
PSO or IPSO	initial population size: 100 inertia weight: 1 inertia weight damping Ratio: 0.99 personal learning coefficient: 1.5 global learning coefficient: 2.0 stopping condition: iterations >1 00 or $\xi < 1 \times 10^{-4}$

The simulation results with various algorithms are given in Table 3. It can be concluded that the integrated method, i.e., IGA and IPSO, had wider applicability and higher accuracy than the traditional GA and PSO. Especially for IGA, it could be used to solve the very small size flute-grinding problem. For further study, the grinding errors ξ (calculated by Equation (14)) between the grinding flute parameters and the designed flute parameters were calculated as shown in Figure 7a. It can be seen that the integrated method superior to the traditional in accuracy and stability. The accuracy of flute parameters with IGA and IPSO could achieve 1×10^{-4}. It was also noted that the grinding errors with IGA and IPSO were less for the medium size flutes, while larger for the small and large size flutes. What is more, to further test the efficiency of the proposed method, the computing time was recorded as shown in Figure 7b. The integrated method also showed better efficiency in convergence, which could save about 15%–40% computing time. In addition, the initial setting parameters of the grinding wheel solved by IGA are provided in Table 4. In light of the IGA solution, three instances were selected and simulated in the software CATIA. The simulated results were obtained and measured, as shown in Figure 8, which also shows that they are highly consistent with the designed flute parameters. In summary, according to the simulation tests for various optimization methods, it is demonstrated that the proposed IGA and IPSO based on the projection model is effective, efficient, and robust solving the flute-grinding problem.

Table 3. The machined flute parameters.

Flute Size	Case No.	Cutter Radius	Desired Flute Parameters [1]	Flute Parameters Solved by IGA	Flute Parameters Solved by GA	Flute Parameters Solved by IPSO	Flute Parameters Solved by PSO
small size flutes	F1	0.3	(0.2, 6, 75)	(0.200, 5.995, 74.956)	/	/	/
	F2	0.5	(0.3, 6, 75)	(0.300, 5.996, 74.976)	/	(0.300, 5.919, 74.796)	/
	F3	1	(0.6, 6, 75)	(0.600, 5.998, 75.962)	(0.602, 5.991, 75.386)	(0.600, 6.000, 75.027)	/
medium size flutes	F4	7	(5, 9, 75)	(5.000, 9.005, 75.016)	(4.989, 8.963, 75.007)	(5.000, 9.000, 74.997)	(5.021, 8.985, 74.781)
	F5	9	(5, 9, 75)	(5.000, 8.996,75.009)	(5.149, 8.894, 75.410)	(5.000, 9.000,75.000)	(4.973, 8.969, 74.926)
	F6	11	(6, 25, 110)	(6.000, 25.016, 109.981)	(5.955, 25.037, 109.530)	(6.000, 25.000,110.002)	(5.946, 24.994, 109.591)
	F7	17	(10, 9, 75)	(10.000, 9.002, 75.005)	(9.997, 9.006, 74.916)	(10.000, 8.998,75.003)	(9.995, 8.984,74.989)
large size flutes	F8	20	(15, 9, 75)	(15.000, 8.995, 75.942)	(14.934, 9.038, 75.343)	(15.000, 9.000, 75.003)	(14.979, 8.995, 75.139)
	F9	25	(17, 25, 110)	(17.000, 24.995, 110.143)	(17.081, 25.071, 109.598)	(17.000, 25.000, 109.998)	(16.929, 24.919, 110.406)
	F10	30	(20, 9, 75)	(20.000, 9.008, 75.047)	(20.129, 9.020, 75.399)	(20.000, 9.000, 74.999)	(20.018, 9.011, 75.109

[1] Note: (core radius, rake angle and flute angle).

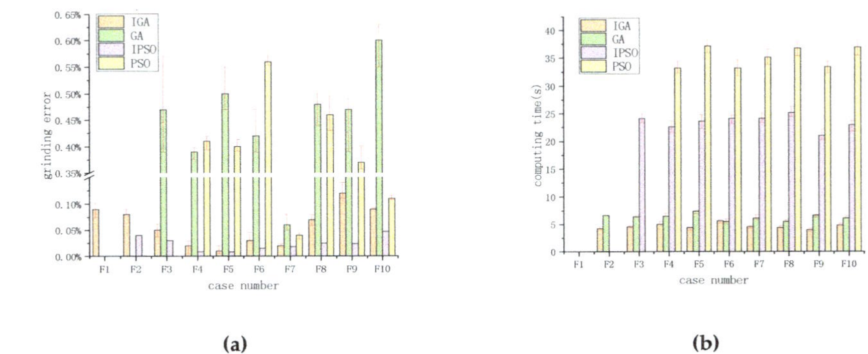

(a)

(b)

Figure 7. Comparisons with various algorithms: (**a**) the grinding errors; (**b**) the computing time.

Table 4. Initial setting parameters of grinding wheels calculated by IGA.

Case No.	β	θ_c	dx	dy
F1	52.9353	87.7783	0.4304	30.1917
F2	50.3040	82.2274	1.7030	30.1838
F3	49.6645	82.7015	1.6804	30.4926
F4	53.2624	95.5550	−3.0894	80.4462
F5	48.4607	80.5059	6.3067	79.4745
F6	55.3177	100.8935	−5.7797	80.4662
F7	48.4607	77.1924	9.5815	83.9187
F8	54.5717	85.3600	3.2566	89.8680
F9	57.7373	99.4303	−6.3210	91.4772
F10	52.8445	79.5509	8.6418	94.2074

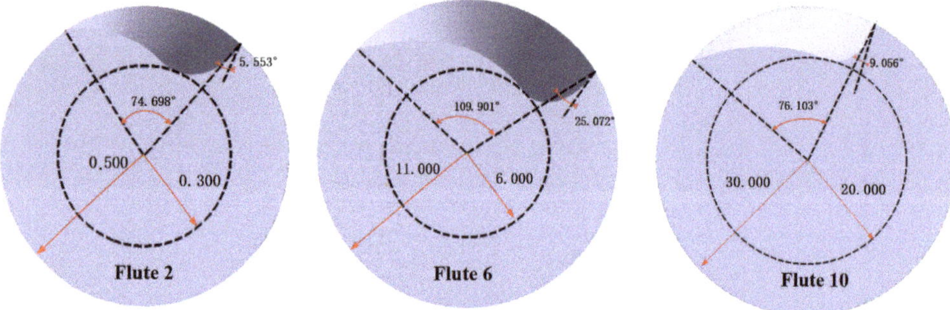

Figure 8. Simulated cutting tool flutes by CATIA.

5. Conclusions

In the CNC flute-grinding processes, the accuracy of generated flute parameters is determined by setting the wheel's location and orientation. The existing methods for the solution of wheel path optimization were time-consuming and cannot handle grinding the extreme-size cutters. In addition, the current model ignored the definition of various constraints in machining, which would strongly affect the machining accuracy and calculation stability.

In the present work, a novel projection model for flute-grinding operations was developed to generate the grinding wheels' configuration. Based on the projection model, the wheel's location was re-formulated with the projection parameters, which simplified the following calculation of machined flute parameters. To minimize the flute-grinding errors regarding the wheel configuration, the projection model was used to generate the proper initial points and was integrated with the GA and PSO as a heuristic regulation. In the numerical simulation, the improved GA and PSO are more accurate, efficient, and robust, with a wide range of applications for various flute sizes. It is noted that the proposed flute-grinding algorithm was verified with a simulation-based method. For the actual grinding, a great deal of topics, such as the dynamics of the grinding machine, the material of the grinding wheel and work-piece, the post-processing, the grinding speed, etc., should be considered in the future experiments.

Author Contributions: Conceptualization, Y.F., L.W. and J.Y.; formal analysis, Y.F., L.W. and J.Y.; funding acquisition, L.W.; Methodology, Y.F., and L.W.; supervision, L.W. and J.L.; Validation, Y.F. and L.W.; writing—original draft, Y.F., L.W. and J.Y.; writing—review and editing, Y.F., L.W. and J.Y. All authors have read and agreed to the published version of the manuscript.

Funding: This work was partially supported by the Funds of the Key Research and Development Plan of Shandong Province (2019GSF108005), the Shandong Provincial Natural Science Foundation, China (ZR2017BEE018) and China Postdoctoral Science Foundation (2016M592182).

Acknowledgments: Our deepest gratitude goes to the editors and the anonymous reviewers for their careful work and thoughtful suggestions that have helped improve this paper substantially.

Conflicts of Interest: The authors declare no conflict of interest.

Nomenclature

R	Grinding wheel radius
$[dx\ dy\ dz]$	Grinding wheel location
β	Grinding wheel orientation
$\mathbf{O_T}$	Tool coordinate system
$\mathbf{O_G}$	Wheel coordinate system
$\mathbf{{}^T M_1}$	Set-up operation matrix
$\mathbf{{}^T M_2}$	Kinematics matrix of 5-axis grinding
v	Translation velocity
ω	Rotation velocity
r_T	Cutter radius
r_c	Core radius
γ	Rake angle
ϕ	Flute angle
r_{c0}	Designed cutter radius
γ_0	Designed flute rake angle
ϕ_0	Designed flute angle
λ	Helix angle

Appendix A. Representation of the Wheel's Location in the Projection Model

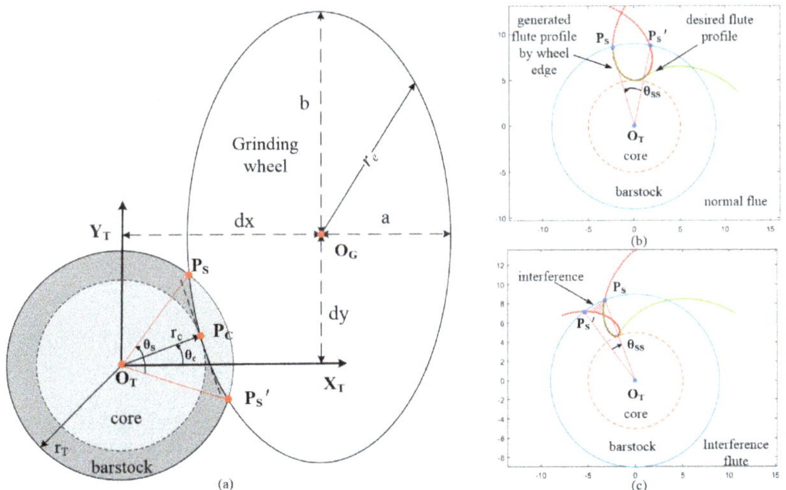

Figure A1. Flute-grinding projected in the cross-section: (**a**) the projection model; (**b**) normal flute; (**c**) interference flute.

In the projection model, the wheel edge can be represented as:

$$r_e = \begin{bmatrix} dx + R \cdot \cos\beta \cdot \cos\theta \\ dy + R \cdot \sin\theta \end{bmatrix}, \tag{A1}$$

The wheel edge is tangent with the core at the point P_c, which satisfies the following condition:

$$\begin{cases} r_c = r_e \\ r_c \cdot r'_e = 0 \end{cases},$$

(A2)

where $r_c = \begin{bmatrix} r_{c0} \cdot \cos\theta_c \\ r_{c0} \cdot \sin\theta_c \end{bmatrix}$ and r'_e is the derivative of r_e.

Solving Equation (A2), we get:

$$\begin{bmatrix} dx \\ dy \end{bmatrix} = \begin{bmatrix} r_{c0} \cdot \cos\theta_c - R \cdot \cos\beta \cdot \cos\theta_e \\ r_{c0} \cdot \sin\theta_c - R \cdot \sin\theta_e \end{bmatrix},$$

(A3)

where $\theta_e = \arctan\left(\frac{\tan\theta_c}{\cos\beta}\right)$.

In summary, the wheel's location $[dx\ dy]$ can be expressed regarding the parameter θ_c.

In addition, two key points in P_S and P'_S in the projection mode were deduced in the following:

$$P_S = \begin{bmatrix} dx + R \cdot \cos\beta \cdot \cos\theta_2 \\ dy + R \cdot \sin\theta_2 \end{bmatrix},$$

(A4)

$$P'_S = \begin{bmatrix} dx + R \cdot \cos\beta \cdot \cos\theta_2 \\ dy + R \cdot \sin\theta_2 \end{bmatrix}.$$

(A5)

The open angle between P_S and P'_S can be represented as: $|O_T P_S| = |O_T P'_S| = r_T$.

Appendix B. The Geometrical Condition for Interference-Free

Improper wheel setting would result in the interference in the flute-grinding operations. As shown in Figure A, the interference is generally caused by the wheel edge grinding in the rake face. Geometrically, the interference would happen while the point P'_S crosses P_S in the counterclockwise direction. To simplify this problem, the angle θ_{ss} is introduced in Figure Ab and c, which is defined as follows:

$$\theta_{ss}\begin{cases} > 0 & No\ interference \\ = 0 & Critical \\ < 0 & Interference \end{cases}.$$

(A6)

In the flute-grinding operations, the angle θ_{ss} can be obtained by mapping the angle θ_s with a phase difference in the projection model: $\theta_{ss} = \theta_s - \Delta\Omega$.

As mentioned, in the flute-grinding process, the tool-stock rotates with speed ω, while the wheel translates in speed v along the Z_T direction. Therefore, the points P_S and P'_S would be located in the cross-section with a phase difference, which can be calculated by the following equation:

$$\Delta\Omega = \frac{f \cdot \sin\beta \cdot (\cos\theta_2 - \cos\theta_1)}{v} \cdot \omega.$$

(A7)

Therefore, to avoid interference, the following constraint can be got for the open-angle θ_s in the projection mode:

$$\theta_s \geq \frac{f \cdot \sin\beta \cdot (\cos\theta_2 - \cos\theta_1)}{v} \cdot \omega.$$

(A8)

References

1. Jiang, F.; Zhang, T.; Yan, L. Analytical model of milling forces based on time-variant sculptured shear surface. *Int. J. Mech. Sci.* **2016**, *115–116*, 190–201. [CrossRef]
2. Yan, L.; Rong, Y.M.; Jiang, F.; Zhou, Z.X. Three-dimension surface characterization of grinding wheel using white light interferometer. *Int. J. Adv. Manuf. Tech.* **2011**, *55*, 133–141. [CrossRef]
3. Pimenov, D.; Hassui, A.; Wojciechowski, S.; Mia, M.; Magri, A.; Suyama, D.; Bustillo, A.; Krolczyk, G.; Gupta, M. Effect of the Relative Position of the Face Milling Tool towards the Workpiece on Machined Surface Roughness and Milling Dynamics. *Appl. Sci.* **2019**, *9*, 842. [CrossRef]
4. Mei, Y.; Mo, R.; Sun, H.; He, B.; Bu, K. Stability Analysis of Milling Process with Multiple Delays. *Appl. Sci.* **2020**, *10*, 3646. [CrossRef]

5. Ren, L.; Wang, S.; Yi, L.; Sun, S. An accurate method for five-axis flute grinding in cylindrical end-mills using standard 1V1/1A1 grinding wheels. *Precis. Eng.* **2016**, *43*, 387–394. [CrossRef]

6. Xiao, S.; Wang, L.; Chen, Z.C.; Wang, S.; Tan, A. A New and Accurate Mathematical Model for Computer Numerically Controlled Programming of 4Y1 Wheels in 21/2-Axis Flute Grinding of Cylindrical End-Mills. *J. Manuf. Sci. Eng.* **2013**, *135*, 04100801–04100811. [CrossRef]

7. Pham, T.T.; Ko, S.L. A manufacturing model of an end mill using a five-axis CNC grinding machine. *Int. J. Adv. Manuf. Tech.* **2010**, *48*, 461–472. [CrossRef]

8. Li, G.; Sun, J.; Li, J. Process modeling of end mill groove machining based on Boolean method. *Int. J. Adv. Manuf. Tech.* **2014**, *75*, 959–966. [CrossRef]

9. Liu, G.; Wei, W.; Dong, X.; Rui, C.; Liu, P.; Li, H. Relief grinding of planar double-enveloping worm gear hob using a four-axis CNC grinding machine. *Int. J. Adv. Manuf. Tech.* **2017**, *89*, 3631–3640. [CrossRef]

10. Van-Hien, N.; Ko, S. A New Method for Determination of Wheel Location in Machining Helical Flute of End Mill. *J. Manuf. Sci. Eng.* **2016**, *138*, 11100301–11100311.

11. Kim, J.H.; Park, J.W.; Ko, T.J. End mill design and machining via cutting simulation. *Comput. Aided Design* **2008**, *40*, 324–333. [CrossRef]

12. Li, G. A new algorithm to solve the grinding wheel profile for end mill groove machining. *Int. J. Adv. Manuf. Tech.* **2017**, *90*, 775–784. [CrossRef]

13. Habibi, M.; Chen, Z.C. A Generic and Efficient Approach to Determining Locations and Orientations of Complex Standard and Worn Wheels for Cutter Flute Grinding Using Characteristics of Virtual Grinding Curves. *J. Manuf. Sci. Eng.* **2017**, *139*, 04101801–04101811. [CrossRef]

14. Wasif, M.; Iqbal, S.A.; Ahmed, A.; Tufail, M.; Rababah, M. Optimization of simplified grinding wheel geometry for the accurate generation of end-mill cutters using the five-axis CNC grinding process. *Int. J. Adv. Manuf. Technol.* **2019**, *105*, 4325–4344. [CrossRef]

15. Kang, S.K.; Ehmann, K.F.; Lin, C. A CAD approach to helical groove machining .1. Mathematical model and model solution. *Int. J. Mach. Tool. Manuf.* **1996**, *36*, 141–153. [CrossRef]

16. Chen, Z.; Ji, W.; He, G.; Liu, X.; Wang, L.; Rong, Y. Iteration based calculation of position and orientation of grinding wheel for solid cutting tool flute grinding. *J. Manuf. Process* **2018**, *36*, 209–215. [CrossRef]

17. Karpuschewski, B.; Jandecka, K.; Mourek, D. Automatic search for wheel position in flute grinding of cutting tools. *Cirp Ann. Manuf. Tech.* **2011**, *60*, 347–350. [CrossRef]

18. Li, G.; Zhou, H.; Jing, X.; Tian, G.; Li, L. An intelligent wheel position searching algorithm for cutting tool grooves with diverse machining precision requirements. *Int. J. Mach. Tool. Manuf.* **2017**, *122*, 149–160. [CrossRef]

19. Li, G.; Zhou, H.; Jing, X.; Tian, G.; Li, L. Modeling of integral cutting tool grooves using envelope theory and numerical methods. *Int. J. Adv. Manuf. Tech.* **2018**, *98*, 579–591. [CrossRef]

20. Wang, L.; Kong, L.; Li, J.; Chen, Z. A parametric and accurate CAD model of flat end mills based on its grinding operations. *Int. J. Precis. Eng. Manuf.* **2017**, *18*, 1363–1370. [CrossRef]

21. Wang, L.; Chen, Z.C.; Li, J.; Sun, J. A novel approach to determination of wheel position and orientation for five-axis CNC flute grinding of end mills. *Int. J. Adv. Manuf. Tech.* **2016**, *84*, 2499–2514. [CrossRef]

Article

Process Parameters Optimization of Thin-Wall Machining for Wire Arc Additive Manufactured Parts

Niccolò Grossi *, Antonio Scippa, Giuseppe Venturini and Gianni Campatelli

Department of Industrial Engineering, University of Firenze, 50139 Firenze, Italy; antonio.scippa@unifi.it (A.S.); giuseppe.venturini@unifi.it (G.V.); gianni.campatelli@unifi.it (G.C.)
* Correspondence: niccolo.grossi@unifi.it

Received: 14 October 2020; Accepted: 23 October 2020; Published: 27 October 2020

Featured Application: This work aims at supporting process parameter selection for machining thin-walled components made by additive manufacturing. Machining industries, especially the ones performing both additive and milling, could benefit from the potential application of such an approach.

Abstract: Additive manufacturing (AM) is an arising production process due to the possibility to produce monolithic components with complex shapes with one single process and without the need for special tooling. AM-produced parts still often require a machining phase, since their surface finish is not compliant with the strict requirements of the most advanced markets, such as aerospace, energy, and defense. Since reduced weight is a key requirement for these parts, they feature thin walls and webs, usually characterized by low stiffness, requiring the usage of low productivity machining parameters. The idea of this paper is to set up an approach which is able to predict the dynamics of a thin-walled part produced using AM. The knowledge of the workpiece dynamics evolution throughout the machining process can be used to carry out cutting parameter optimization with different objectives (e.g., chatter avoidance, force vibrations reduction). The developed approach exploits finite element (FE) analysis to predict the workpiece dynamics during the machining process, updating its changing geometry. The developed solution can automatically optimize the toolpath for the machining operation, generated by any Computer Aided Manufacturing (CAM) software updating spindle speed in accordance with the selected optimization strategies. The developed approach was tested using as a test case an airfoil.

Keywords: additive manufacturing; thin walled machining; dynamics; machining cycle optimization

---

1. Introduction

Lightweight constructions are becoming more and more important for many industries, such as power generation, aerospace, automotive and medical technology. The functional requirements of these parts often lead to the inclusion of thin-walled features in the design of such components. Due to the strict strength and fatigue requirements, such structures are often machined from the bulk as monolithic components, removing a large amount of material to create the final product. Usually this is achieved through intensive milling operations, removing up to 95% of the stock volume, to create the final geometry [1].

An arising trend in thin-walled parts manufacturing is the introduction of additive manufacturing (AM) in the process chain. Indeed, AM allows one to deposit only the material needed for the thin-walled structures, considerably reducing the material waste. This is a critical issue, especially when high cost materials are used, such as titanium or nickel-based alloys. Nowadays, many different AM processes are available on the market [2]: powder bed fusion processes such as electron beam melting or selective laser melting enable the manufacturing of extremely complex geometries, but have strict limitations on the part size [3]; on the other hand, directed energy deposition processes such as laser deposition [4] or wire-arc-additive-manufacturing (WAAM) [5] enable the manufacturing of large parts with a limited features resolution. Despite the great advancements undergone by the AM processes in terms of material quality [6] and repeatability, these technologies often do not meet the dimensional and surface finishing requirements prescribed by the applications [7]. For this reason, milling operations are usually carried out after AM, to correct inaccuracies and improve the surface quality [8]. Therefore, including AM in the process chain requires carrying out machining of thin-walled components. This is a critical operation due to the low stiffness of such workpieces, making them prone to forced vibrations, chatter [9], and deflection issues [10], being responsible for surface location errors [11] and poor surface finish [12,13].

The general approach used to prevent these issues is to limit the material removal rate (MRR) by means of conservative machining parameters. This generates a reduction in the cutting forces, hence limiting workpieces' vibrations at the expense of reduced productivity. Counteracting process vibrations without affecting the productivity requires the knowledge of workpiece dynamic behavior, i.e., the frequency response functions (FRFs), at each driving point triggered during the milling cycle [14,15]. This is hardly practicable through experimental techniques, such as impact testing, since the workpiece FRFs change during the milling operation, due to the material removal process [16]. Moreover, in a thin-walled workpiece, the driving points' FRFs are strongly dependent on the excitation point, resulting in a different dynamic behavior over the component. Finite element (FE) models are a convenient way of overcoming these issues, enabling a virtual identification of workpiece dynamic behavior at different driving points [17].

This paper proposes an innovative approach to identify the dynamic behavior of thin-walled workpieces during milling operations. The basis of the proposed technique is the identification of the workpiece dynamics through FE modelling using 2D shell elements [18]. This approach enables an efficient and accurate description of thin-walled structures' dynamics. Moreover, the generation of a shell model could be easily automated exploiting AM deposition path, since it provides the information concerning the workpiece skeleton surface.

The proposed technique is divided into three main stages: (i) AM stock modelling, (ii) stock thickness updating and (iii) FRF identification. In the first step, a shell FE model of the deposed material is created. Then, the machine tool numerical control (NC) programming language, known as G-code, is analyzed through an automated algorithm that identifies the position of the tool–workpiece contact point and updates the stock geometry, modifying the shell element's thickness. This information allows the system to generate an updated FE model of the workpiece that takes the material removal effect into account, enabling the accurate identification of its dynamic behavior. In the final step, the workpiece FRFs are calculated at the driving points (i.e., the tool–workpiece contact point) for every step of the machining process. This provides all the required information to optimize the milling process.

To prove the effectiveness and accuracy of the proposed dynamics identification technique, two specimens of a test case thin-walled component were manufactured through the WAAM process. In this work, the WAAM process was selected because its high productivity has the drawback of low accuracy and the need for an extensive subtractive process to realize the final product. However, the proposed approach could be applied to other directed energy deposition processes. A five-axis milling cycle was defined and analyzed using the proposed technique, identifying the evolution of workpiece dynamics. Then, the machining parameters (i.e., spindle speed and feed rate) were adjusted following two different optimization techniques. The two specimens were machined, interrupting the process to perform modal analysis, required to verify the accuracy of the proposed modelling technique. At the end of the milling process, both specimens were analyzed through surface measurements.

2. Additive Production of Thin Walled Component

The selected test case is an National Advisory Committee for Aeronautics 9403 airfoil, (NACA, USA) as shown in Figure 1. The stock to be machined was initially created using WAAM (Figure 1a) and a subsequent 5-axis milling step was used to obtain the final shape (in Figure 1b). In WAAM, subsequent layers of metal are selectively deposed using a gas metal arc-welding source, in which raw material in the form of wire is molten through an electric arc. This enables one to achieve high productivity and to manufacture very large components if compared with alternative AM processes [19].

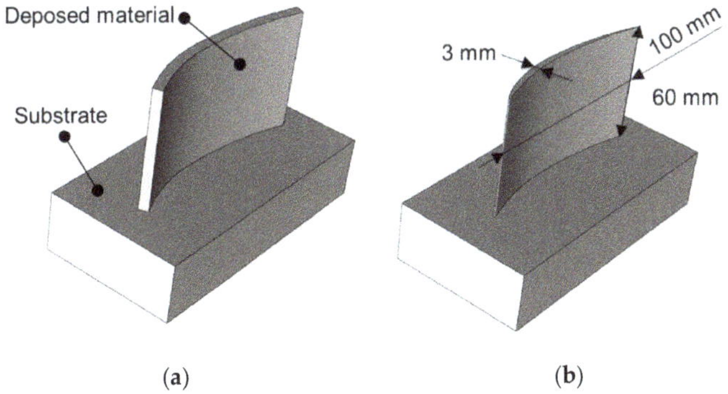

Figure 1. The selected test case after the deposition (**a**) and the milling process (**b**).

The airfoil was created deposing 41 layers of ER70S-6 (i.e., a standard filler metal for carbon steels) onto a brick-shaped substrate made of AISI 1040. The deposition was carried out using the process parameters presented in Table 1.

Table 1. Welding parameters used in the deposition.

Current (A)	Voltage (V)	Deposition Speed (mm/min)	Wire Feed Speed (m/min)
80	20	200	4.6

Such parameters resulted in an average layer height of 1.8 mm and in a layer width of about 6.8 mm and have been selected based on previous studies [20]. The filler material was deposed following the airfoil camber line, creating a constant cross-sectional profile. The final shape of airfoil was achieved by the subsequent 5-axis milling process. The WAAM machine prototype developed by the Manufacturing technologies research laboratory (MTRL) of the University of Firenze was used to carry out the deposition [21]. The deposition process is shown in Figure 2a, while Figure 2b shows the airfoil after the WAAM step. Figure 2b highlights that a significant surface waviness issue affects the WAAM component, requiring a milling step, performed on Mori Seiki (Nagoya, Japan) NMV 1500 DCG milling machine, to meet the surface finishing requirement. As stated in the introduction, two specimens of the selected test case were manufactured. After the deposition step, the geometry of the WAAM parts was acquired through a Euro Apex C776 coordinates-measurement-machine (CMM), (Mitutoyo, Japan). This step allowed us to check the required machining allowances for the milling process. The CMM measurement operation is shown in Figure 3a, while Figure 3b shows the CMM measurements results. CMM measurements highlighted that the airfoil surfaces presented an average deviation from the reference surfaces of about 0.24 mm.

(**a**) (**b**)

Figure 2. Wire-arc-additive-manufacturing (WAAM) deposition (**a**), airfoil after the deposition (**b**).

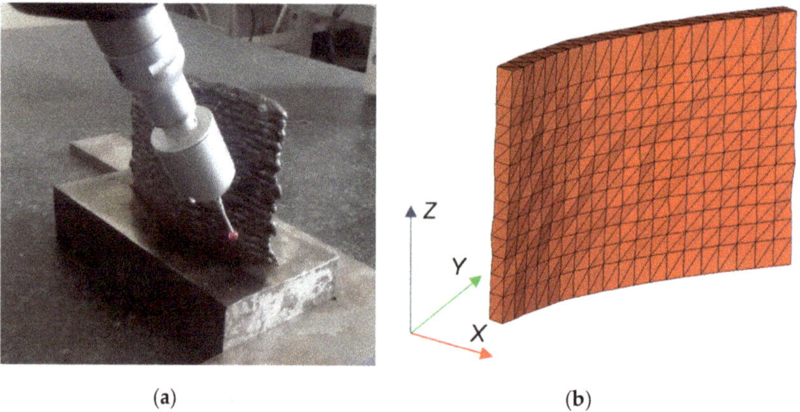

(**a**) (**b**)

Figure 3. Coordinates-measurement-machine (CMM) measurement of the deposed airfoil (**a**), measured airfoil geometry (**b**).

3. Adopted Machining Cycle

The machining cycle adopted to produce the airfoil requires 5-axis machining capabilities to ensure the manufacturability of the part. The cycle is constituted by three phases: roughing, semi-finishing and finishing. The tool used for all the operations is a 2-flute 10 mm carbide ball end mill produced by cod. 207280 (Garant, Lengerich, Germany) designed for dry cutting condition on carbon steel. A morphing strategy was applied for the roughing phase to reach an offset geometry of the airfoil starting from the nearly constant thickness geometry produced by WAAM process. A 2 mm axial depth of cut was set for this operation. In Figure 4, the roughing operation is presented. The following machining parameters were selected, based on previous experiences of WAAM parts machining [22]: cutting speed 180 m/min and feed per tooth 0.062 mm/tooth. A machining allowance of 0.5 mm was left on the product, to be removed in the following operations. The Rotation Tool Center Point (RTCP) function was activated for the G-code set-up and a constant 30° tilt angle of the tool with respect to the airfoil surface was used. As described in Section 2, CMM measurement of the WAAM part was used to optimize the alignment between the final part and the stock. The oriented geometry of the airfoil was then used to create the toolpath thanks to the ESPRIT® (DP Technology, Camarillo, CA, USA) Computer Aided Manufacturing (CAM) software. After the roughing cycle, semi-finishing and finishing operations were carried out to complete the manufacturing of the airfoil. The two cycles used the same tilt angle of the roughing phase. The axial depth of cut adopted for the machining operations was 1 mm for semi-finishing and 0.5 mm for finishing. These steps were selected considering the expected surface finish.

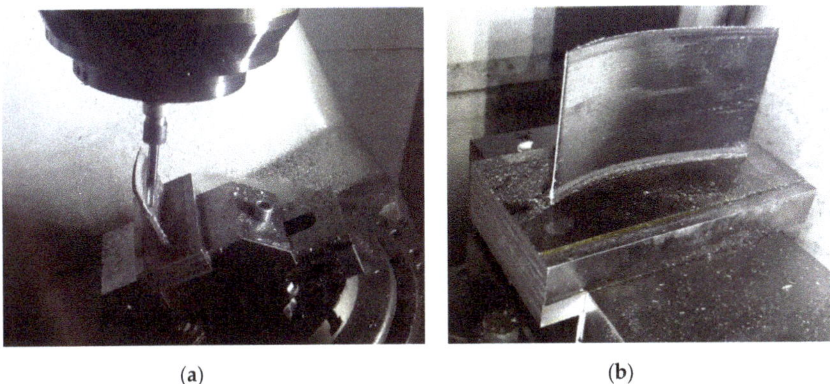

(a)   (b)

Figure 4. (a) Roughing cycle, (b) finished part.

4. Thin-Walled Dynamics Prediction

Thin wall dynamics can be proficiently estimated using finite element analysis. To increase modelling reliability and to reduce the computational effort, shell elements were considered. Moreover, considering the application of the proposed technique to the WAAM process, the deposition path can be used to automatically generate the shell FE mesh. Each point of the NC code generated to create the deposition toolpath can be used to become a node of the FE model, where only the properties and thickness of the shell elements shall be adjusted to create a realistic model. This enables a fast and automatic mesh and model creation starting from the simple G-code of the deposition process (see Figure 5). To take the material removal into account, the FE model must be continuously updated, according to the tool position along to the CAM generated toolpath. For this reason, a general algorithm was developed to modify the nodal thicknesses, according to the tool engagement obtained by a post-processing of the G-code (Figure 6).

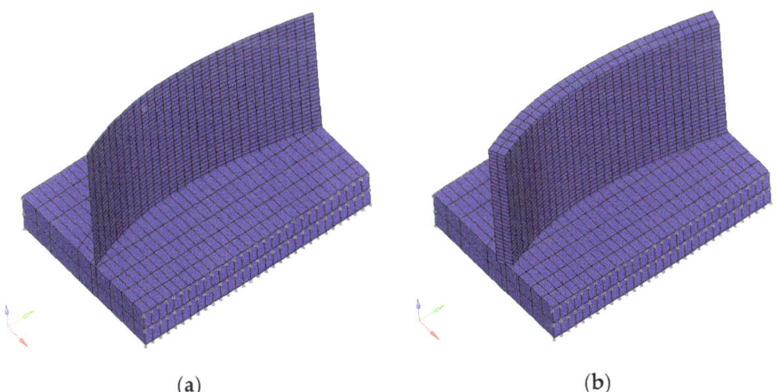

(a)   (b)

Figure 5. (a) Traditional shell element representation, (b) 3D element representation, showing the nodal thickness.

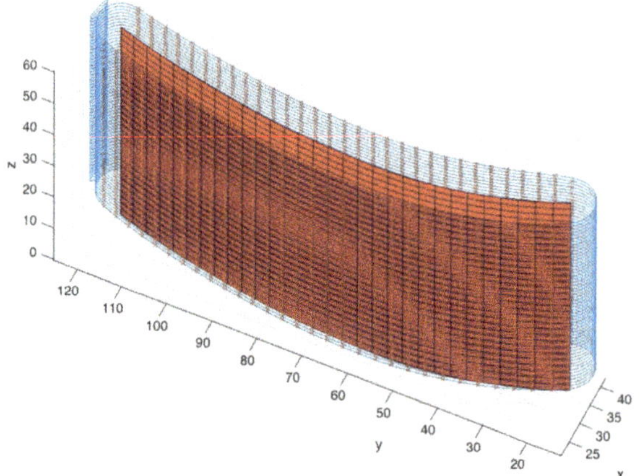

Figure 6. Roughing cycle toolpath post-processing.

For each node of the FE model, the corresponding block of the G-code is determined, computed as the minimum distance between the tooltip and the node considered. The nodal thickness is then updated considering the intersection of the tooltip geometry (a ball nose mill for the considered test case) with the external surface of the machined component (Figure 7). The predicted frequency response (FR) of the machined part can thus be easily obtained, performing a numerical frequency response analysis. For the test case considered, a modal frequency response was preferred. A unit force was applied to the node closest to the tool-tip position (obtained by G-code), directed normal to the surface. Linear behavior was assumed, and dynamic compliance in the other direction was neglected. This allows for obtaining a good description of thin wall dynamics in a wide frequency range using a limited number of excitation frequencies, notably reducing the computational effort: a FR analysis took about 3 s on a PC using a commercial FE solver (MSC Nastran 2014.1, Swedish Hexagon AB, Newport Beach, CA, USA). The accuracy of the FE model was tested after the roughing machining phase, considering the FE model, updated according to the aforementioned algorithm (Figure 8). The following mechanical characteristics were considered: elastic modulus: 200 GPa, density: 7850 kg/m^3, Poisson's ratio: 0.31, structural damping: 0.005. According to the results in Figure 1, the proposed approach allows for predicting the changing of workpiece dynamics along the toolpath, taking the material removal into account at the same time.

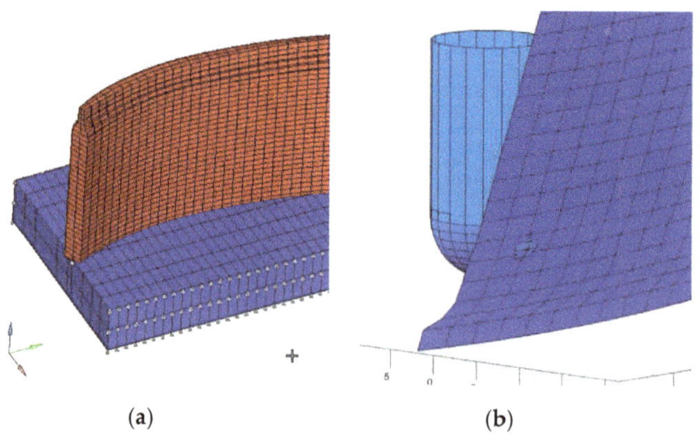

(a) (b)

Figure 7. Finite element (FE) model updating phase: (**a**) FE model (**b**) tooltip position.

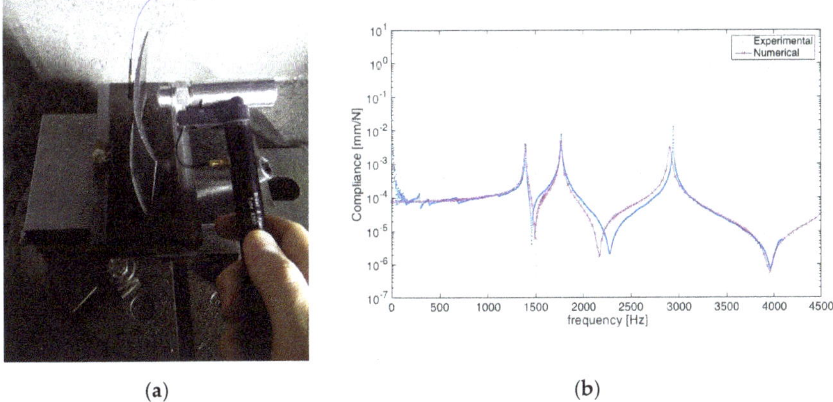

(a)   (b)

Figure 8. Workpiece driving point frequency response functions (FRFs): (**a**) experimental setup (**b**) comparison with numerical FRF.

Simulated workpiece FRFs can be then composed, according with the orientation of the tool (lead and tilt angle) with the tooltip FRFs (experimentally determined), to obtain the dynamics of the system (tool + workpiece), changing with respect to the cutting point (see Figure 9). Combination of tool–workpiece compliance was used in the optimization strategy.

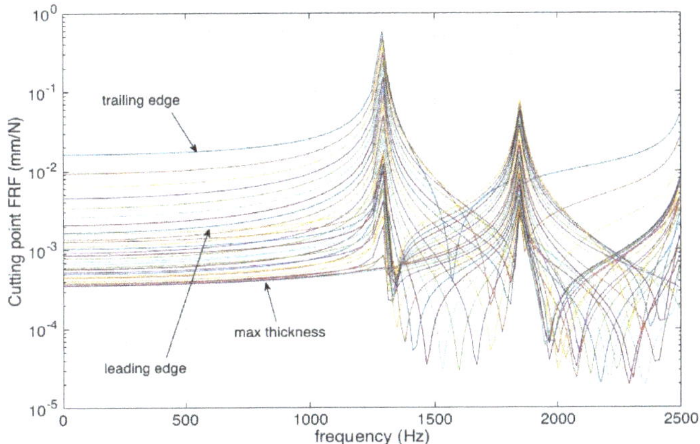

Figure 9. Compliance at the tip of the blade moving from the trailing to the leading edge.

5. Optimization Strategy

The proposed approach allows for identifying the dynamics of the tool–workpiece system during the whole machining cycle and could be coupled with any optimization strategy to achieve different goals, computing an optimized machining cycle.

Two main dynamic phenomena could arise in machining thin-walled components: forced vibrations and chatter vibrations. Both the effects are minimized by increasing the stiffness of the system or reducing the MRR (i.e., depth of cut). The first approach is hard to follow since workpiece shape and tooling system are generally fixed. The second one affects the productivity of the process.

Therefore, optimization strategies are generally based on the selection of spindle speed [23], which governs the frequency content of the cutting forces and thus could reduce vibration levels, without affecting productivity or altering system layout. However, the selection of spindle speed has conflicting requirements for forced and chatter vibrations. Indeed, to reduce forced vibrations, the spindle speed should be far from the resonance frequencies of the system, hence reducing the relative tool–workpiece displacements during the cycle. On the contrary, for chatter avoidance, the optimal spindle speeds excite the resonance frequency of the dominant mode [24].

According to these considerations, the machining cycle optimization strategy should be tailored for the specific application. In this work, two simple optimization strategies were applied, with the aim of showing the potential applications of the proposed approach.

- Strategy A: ensuring chatter-free machining by selecting a spindle speed, exciting the resonance of the dominant mode
- Strategy B: minimizing forced vibrations by selecting spindle speeds far from the resonances of the systems

Both the strategies were applied to the test case, starting from the dynamic behavior of the system at the cutting point computed by the proposed technique, presented in the previous section.

For strategy A, the dominant mode of the system was identified in the different steps of the machining cycle and a single optimized spindle speed was selected, ensuring the excitation of the dominant mode with one of the tooth pass frequency (f_{tp}) harmonics. The procedure is exemplified in Figure 10, where FRFs of system during the machining cycle at the trailing edge are presented. The figure shows the dominant mode of the system around 1300 Hz. A single spindle speed equal to 5580 rpm (f_{tp} = 186 Hz) was selected to ensure the system to work close to the dominant mode resonance frequency with one of the f_{tp} harmonics (7th).

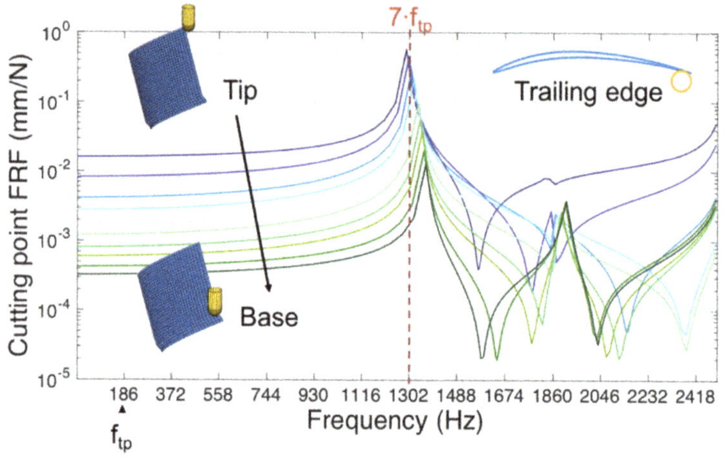

Figure 10. Optimization strategy A applied to the airfoil.

For strategy B, machining parameters were selected to minimize the relative displacement of the tool–workpiece under the effect of cutting force (i.e., force vibrations). These vibrations were assessed by computing the system compliance at the f_{tp} and harmonics (till 12th).

The harmonics were weighted based on the results of a preliminary analysis in which cutting forces were simulated in the finishing cutting conditions (i.e., radial depth of cut (*ae*) 0.2 mm, axial depth of cut (*ap*) 0.5 mm).Fast Fourier Transform (FFTs) of the cutting force and weights adopted are presented in Figure 11. This approach allows for identifying the relative displacement of the tool and workpiece in the case of stable cutting (i.e., without chatter) and, based on this information, selecting the optimal speed in a defined range (in this case, 140 to 235 m/min). The interesting advantage of such approach is not considering a cutting force model and the consequently need of cutting force coefficients [25]—this allows us to perform the proposed optimization but not to predict the actual dynamic deflection.

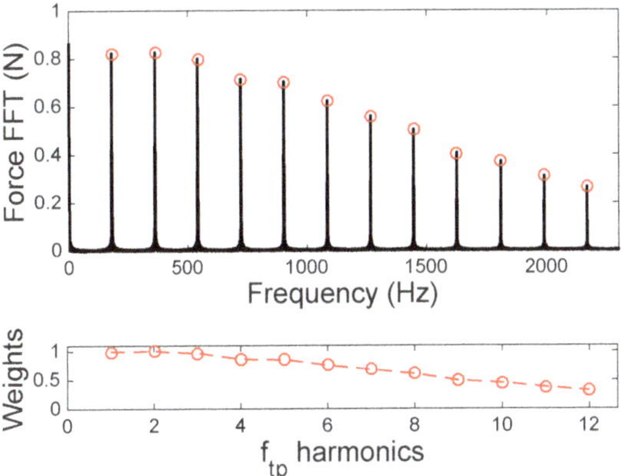

Figure 11. Harmonics weights for optimization strategy B.

To effectively minimize the forced vibrations, this procedure was repeated for each step of the cycle, and with respect to the previous strategy, the spindle speed was changed continuously during the machining cycle. Feed rate was changed accordingly to keep feed per tooth constant (0.062 mm/tooth). The optimal cutting speeds trend for the finishing of the airfoil, using strategy B, is presented in Figure 12.

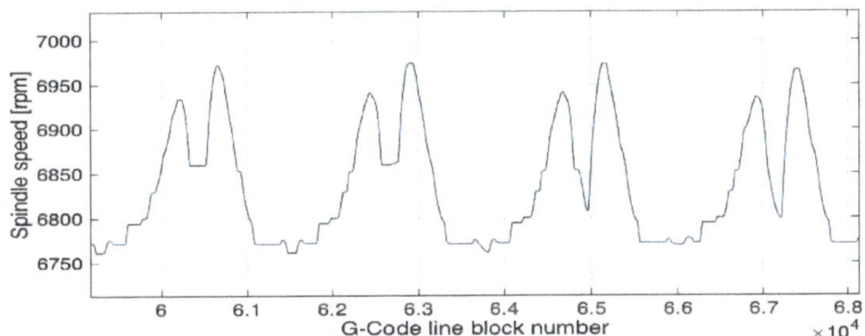

Figure 12. Optimal spindle speeds for optimization strategy B.

6. Results and Discussion

The two strategies were tested on the finishing phase of the test-case to assess their performances. During strategy B machining, chatter occurred, as was clear from the sound during cutting and the surface quality of the product. Strategy B was defined to minimize forced vibrations assuming a chatter-free condition; however, adopting such a solution leads to the selection of spindle speeds more prone to chatter vibrations, as explained in the previous section. On the contrary, no chatter was detected during the strategy A machining cycle: the selection of the suitable spindle speed avoided the occurrence of the phenomenon.

To further investigate the results, airfoils surfaces were acquiring by the CMM and analyzed. Both pressure and suction sides were scanned using a 2 mm spatial step along length and height. First, form errors of the surfaces respect to the nominal NACA 9403 were studied. Mid-surfaces were computed and compared with the reference camber line and results are shown at three heights (Figure 13): Tip is close to the upper part of the airfoil, Mid is in the middle and Base is close to the fixed root part but higher because of measurement limits. Both strategies could accurately reproduce the camber line with very small deviations with respect to the reference (less than 0.1 mm). Indeed, this behavior is in line with the model results: almost the same level of vibrations in the same point of the camber line on the two sides of the airfoil (pressure and suction) was predicted, and therefore no significant modifications of the airfoil camber line were expected.

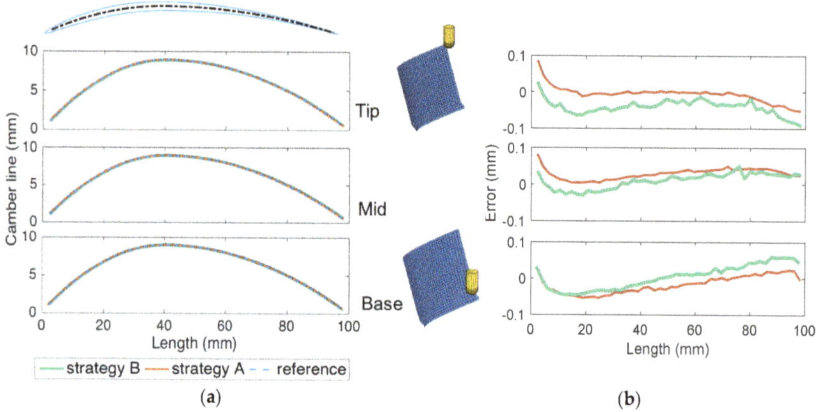

Figure 13. Camber line at different Z-level (**a**) camber line (**b**) camber line error.

On the contrary, significant thickness differences are expected along the camber line, caused by static deflection and forced vibrations. To assess this effect, the thickness of the product was computed at the different levels and compared to the desired one Figure 14. The results show a higher thickness at the borders, especially at the trailing edge, increasing from the base to the tip. This is due to compliance of the system, causing workpiece deflection under the cutting force effect. This deflection increases where tool–workpiece relative displacements are higher, i.e., at the trailing and leading edges and at the tip, producing an increased thickness. Moreover, in Figure 14 on the right, errors on both sides are shown, highlighting similar errors between the two strategies and the same level of error on both sides, as predicted.

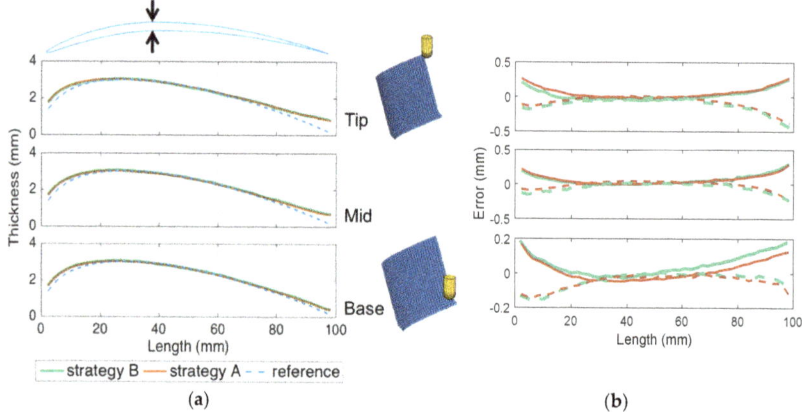

Figure 14. Thickness at different Z-level (**a**) airfoil thickness (**b**) error on suction (solid line) and pressure (dashed line) sides.

The thickness error along the entire surface is presented in Figures 15 and 16. The results present the thickness error trend, higher at the two edges and increasing the height. It is interesting to point out that the thickness error is very similar between the two strategies. This could be explained by the fact that strategy A allowed us to avoid chatter vibrations at the price of working at the resonance of the system, thus increasing forced vibrations and causing thickness errors. Strategy B was selected to reduce forced vibrations, but chatter vibration occurred, leading to a significant increase in cutting forces, which cancelled the optimized speed selection benefits. Moreover, static deflection is not affected by the two strategies and could be the main factor responsible for the deviations that increase at the leading and trailing edge because of structural stiffness reduction. The main difference between the two thickness error maps is roughness: strategy A produced a smoother surface, while strategy B presents higher waviness and roughness caused by the occurrence of chatter.

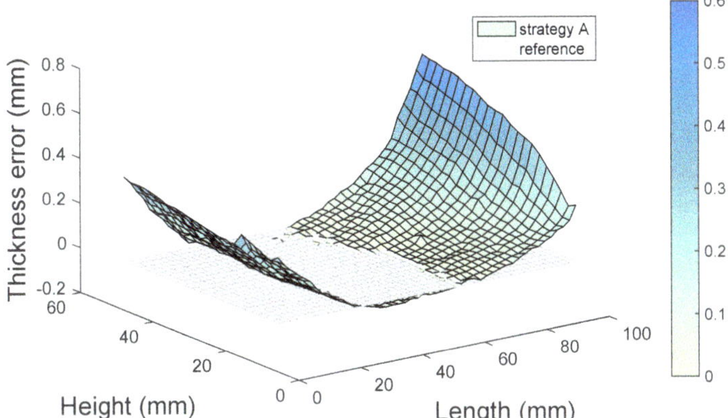

Figure 15. Airfoil thickness error for strategy A.

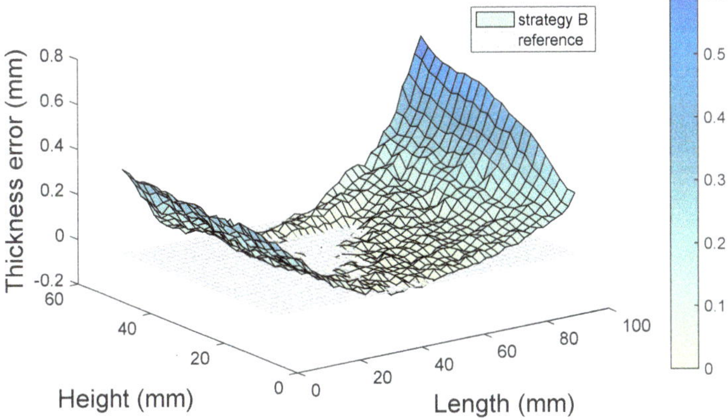

Figure 16. Airfoil thickness for strategy B.

7. Conclusions

In this paper, a novel integrated approach to predict the changing dynamic behavior of a thin walled component is presented. The thin-walled test case, a NACA 9403 airfoil, was initially created using WAAM, an arising technology to produce low buy-to-fly ratio components. The developed system was able to correctly simulate both the material removal process on an initial stock and its dynamic behavior during the machining cycle. The results show that the proposed method can be proficiently applied for AM products thanks to the automatic FE mesh generation and updating.

Based on the simulation results, two different optimization strategies were applied, confirming the importance of the right spindle speed selection. In this work, milling of an additive manufactured component has been optimized based only on the dynamic behavior of the thin-walled part. In the future, a more comprehensive approach, including cutting force and tool wear, could be implemented. In that case, the process parameters of the additive manufacturing step should be taken into account since they affect the proprieties of the part [26]. The idea underpinning this work is investigating hybrid additive-subtractive manufacturing; in this work, the approach was applied using to two separated machines, and in the future the combination of milling and WAAM on the same machine will be investigated.

Author Contributions: Methodology and conceptualization, N.G. and A.S., investigation and validation, N.G. and G.V.; data curation and writing—original draft preparation, N.G.; supervision, A.S. and G.C.; project administration and funding acquisition, G.C. All authors have read and agreed to the published version of the manuscript.

Funding: This research received no external funding.

Acknowledgments: The authors would like to thank Machine Tool Technology Research Foundation (MTTRF) and its supporters for the loaned machine tool (Mori Seiki NMV1500DCG) and CAM Software (ESPRIT), both used to carry out this research activity.

Conflicts of Interest: The authors declare no conflict of interest. The funders had no role in the design of the study; in the collection, analyses, or interpretation of data; in the writing of the manuscript, or in the decision to publish the results.

References

1. Oliveira, D.; Calleja, A.; López de Lacalle, L.N.; Campa, F.; Lamikiz, A. Flank milling of Complex Surface. In *Machining of Complex Sculptured Surfaces*; Davim, J.P., Ed.; Springer: London, UK, 2012.
2. Frazier, W.E. Metal Additive Manufacturing: A Review. *J. Mater. Eng. Perform.* **2014**, *23*, 1917–1928. [CrossRef]
3. Sing, S.L.; Yeong, W.Y. Laser powder bed fusion for metal additive manufacturing: Perspectives on recent developments. *Virtual Phys. Prototyp.* **2020**, *15*, 359–370. [CrossRef]
4. Guo, C.; He, S.; Yue, H.; Li, Q.; Hao, G. Prediction modelling and process optimization for forming multi-layer cladding structures with laser directed energy deposition. *Opt. Laser Technol.* **2021**, *134*, 106607, in press. [CrossRef]
5. Montevecchi, F.; Venturini, G.; Grossi, N.; Scippa, A.; Campatelli, G. Heat accumulation prevention in Wire-Arc-Additive-Manufacturing using air jet impingement. *Manuf. Lett.* **2018**, *17*, 14–18. [CrossRef]
6. Wächter, M.; Leicher, M.; Hupka, M.; Leistner, C.; Masendorf, L.; Treutler, K.; Kamper, S.; Esderts, A.; Wesling, V.; Hartmann, S. Monotonic and Fatigue Properties of Steel Material Manufactured by Wire Arc Additive Manufacturing. *Appl. Sci.* **2020**, *10*, 5238. [CrossRef]
7. Li, F.; Chen, S.; Shi, J.; Tian, H.; Zhao, Y. Evaluation and Optimization of a Hybrid Manufacturing Process Combining Wire Arc Additive Manufacturing with Milling for the Fabrication of Stiffened Panels. *Appl. Sci.* **2017**, *7*, 1233. [CrossRef]
8. Flynn, J.M.; Shokrani, A.; Newman, S.T.; Dhokia, V. Hybrid additive and subtractive machine tools—Research and industrial developments. *Int. J. Mach. Tools Manuf.* **2016**, *101*, 79–101. [CrossRef]
9. Puma-Araujo, S.D.; Olvera-Trejo, D.; Martínez-Romero, O.; Urbikain, G.; Elías-Zúñiga, A.; López de Lacalle, L.N. Semi-Active Magnetorheological Damper Device for Chatter Mitigation during Milling of Thin-Floor Components. *Appl. Sci.* **2020**, *10*, 5313. [CrossRef]
10. Grossi, N.; Scippa, A.; Croppi, L.; Morelli, L.; Campatelli, G. Adaptive toolpath for 3-axis milling of thin walled parts. *MM Sci. J.* **2019**, *2019*, 3378–3385. [CrossRef]
11. Ning, H.; Zhigang, W.; Chengyu, J.; Bing, Z. Finite element method analysis and control stratagem for machining deformation of thin-walled components. *J. Mater. Process. Technol.* **2003**, *139*, 332–336. [CrossRef]

12. Lin, Y.-C.; Wu, K.-D.; Shih, W.-C.; Hsu, P.-K.; Hung, J.-P. Prediction of Surface Roughness Based on Cutting Parameters and Machining Vibration in End Milling Using Regression Method and Artificial Neural Network. *Appl. Sci.* **2020**, *10*, 3941. [CrossRef]

13. Grossi, N.; Scippa, A.; Sallese, L.; Montevecchi, F.; Campatelli, G. On the generation of chatter marks in peripheral milling: A spectral interpretation. *Int. J. Mach. Tools Manuf.* **2018**, *133*, 31–46. [CrossRef]

14. Arnaud, L.; Gonzalo, O.; Seguy, S.; Jauregi, H.; Peigné, G. Simulation of low rigidity part machining applied to thin-walled structures. *Int. J. Adv. Manuf. Technol.* **2011**, *54*, 479–488. [CrossRef]

15. Scippa, A.; Grossi, N.; Campatelli, G. FEM based Cutting Velocity Selection for Thin Walled Part Machining. *Procedia CIRP* **2014**, *14*, 287–292. [CrossRef]

16. Thevenot, V.; Arnaud, L.; Dessein, G.; Cazenave-Larroche, G. Influence of Material Removal on the Dynamic Behavior of Thin-walled strucutres in Peripheral Milling. *Mach. Sci. Technol.* **2006**, *10*, 275–287. [CrossRef]

17. Tuysuz, O.; Altintas, Y. Frequency Domain Prediction of Varying Thin-Walled Workpiece Dynamics in Machining. *J. Manuf. Sci. Eng.* **2017**, *139*, 071013. [CrossRef]

18. Zienkiewicz, O.C.; Taylor, R.L. *The Finite Element Method-The Basis*, 5th ed.; Butterworth-Heinemann: Woburn, UK, 2000.

19. Ding, D.; Pan, Z.; Cuiuri, D.; Li, H. Wire-feed additive manufacturing of metal components: Technologies, developments and future interests. *Int. J. Adv. Manuf. Technol.* **2015**, *81*, 465–481. [CrossRef]

20. Campatelli, G.; Campanella, D.; Barcellona, A.; Fratini, L.; Grossi, N.; Ingarao, G. Microstructural, mechanical and energy demand characterization of alternative WAAM techniques for Al-alloy parts production. *CIRP J. Manuf. Sci. Technol.* **2020**, in press. [CrossRef]

21. Montevecchi, F.; Venturini, G.; Scippa, A.; Campatelli, G. Finite Element Modelling of Wire-arc-additive-manufacturing Process. *Procedia CIRP* **2016**, *55*, 109–114. [CrossRef]

22. Montevecchi, F.; Grossi, N.; Takagi, H.; Scippa, A.; Sasahara, H.; Campatelli, G. Cutting forces analysis in additive manufactured AISI H13 alloy. *Procedia CIRP* **2016**, *46*, 476–479. [CrossRef]

23. Bolsunovskiy, S.; Vermel, V.; Gubanov, G.; Kacharava, I.; Kudryashov, A. Thin-Walled Part Machining Process Parameters Optimization based on Finite-Element Modeling of Workpiece Vibrations. *Procedia CIRP* **2013**, *8*, 276–280. [CrossRef]

24. Montevecchi, F.; Grossi, N.; Scippa, A.; Campatelli, G. Improved RCSA technique for efficient tool-tip dynamics prediction. *Precis. Eng.* **2016**, *44*, 152–162. [CrossRef]

25. Grossi, N. Accurate and fast measurement of specific cutting force coefficients changing with spindle speed. *Int. J. Precis. Eng. Manuf.* **2017**, *18*, 1173–1180. [CrossRef]

26. Tan, J.H.K.; Sing, S.L.; Yeong, W.Y. Microstructure modelling for metallic additive manufacturing: A review. *Virtual Psysical Prototyp.* **2020**, *15*, 87–105. [CrossRef]

Publisher's Note: MDPI stays neutral with regard to jurisdictional claims in published maps and institutional affiliations.

 *applied sciences*

Article

A Qualitative Tool Condition Monitoring Framework Using Convolution Neural Network and Transfer Learning

Harshavardhan Mamledesai, Mario A. Soriano and Rafiq Ahmad *

Laboratory of Intelligent Manufacturing, Design and Automation (LIMDA),
Department of Mechanical Engineering, University of Alberta, Edmonton, AB T6G1H9, Canada;
mamledes@ualberta.ca (H.M.); masorian@ualberta.ca (M.A.S.)
* Correspondence: rafiq.ahmad@ualberta.ca

Received: 26 September 2020; Accepted: 16 October 2020; Published: 19 October 2020

Abstract: Tool condition monitoring is one of the classical problems of manufacturing that is yet to see a solution that can be implementable in machine shops around the world. In tool condition monitoring, we are mostly trying to define a tool change policy. This tool change policy would identify a tool that produces a non-conforming part. When the non-conforming part producing tool is identified, it could be changed, and a proactive approach to machining quality that saves resources invested in non-conforming parts would be possible. The existing studies highlight three barriers that need to be addressed before a tool condition monitoring solution can be implemented to carry out tool change decision-making autonomously and independently in machine shops around the world. First, these systems are not flexible enough to include different quality requirements of the machine shops. The existing studies only consider one quality aspect (for example, surface finish), which is difficult to generalize across the different quality requirements like concentricity or burrs on edges commonly seen in machine shops. Second, the studies try to quantify the tool condition, while the question that matters is whether the tool produces a conforming or a non-conforming part. Third, the qualitative answer to whether the tool produces a conforming or a non-conforming part requires a large amount of data to train the predictive models. The proposed model addresses these three barriers using the concepts of computer vision, a convolution neural network (CNN), and transfer learning (TL) to teach the machines how a conforming component-producing tool looks and how a non-conforming component-producing tool looks.

Keywords: tool condition monitoring; tool change policy; Industry 4.0; machine learning; CNN; AI

1. Introduction

The last decade has been the decade of the fourth industrial revolution (Industry 4.0) for manufacturing around the world using smart systems and technology. Industry 4.0 uses the concepts of machine learning and big data to reduce wastage of time and resources and make the production process more efficient [1,2]. The concepts of Industry 4.0 require machines that are smart and autonomous [3], and this presents an excellent opportunity for the development of machine learning algorithms that improve operations and help in the reduction of waste.

Smart systems are essential for the implementation of Industry 4.0 in machine shops, but what is the definition of a smart machine in the context of machining quality control? There is no clear definition proposed in past studies. A smarter machine in the context of machining quality control can be envisioned as a machine with the intelligence to understand and implement the quality requirements. This machine only produces the parts that meet the design requirements (i.e., conforming parts) and detects the changes in machining parameters or environmental factors.

These changes in environmental factors might result in the manufacturing of parts that do not meet the design requirements (i.e., non-conforming parts). In other words, the machine has the intelligence to predict whether the machine will produce a conforming or a non-conforming part based on the environmental inputs.

In a mass manufacturing facility, the environmental factors like type of machine used, jigs, and fixtures in which the machining is taking place are reasonably stable, and these are changed only when the production lines are repurposed to produce different components. Given the stable environmental factors, the quality of the machining largely depends on the consumables like cutting tools, coolant oils, and others used by these machining processes [4,5]. If the consumables are used for too long, they contribute to the production of non-conforming parts, and if the consumables are underutilized, they add to the overheads and wastage [6]. The process of defining the limits for an overused and an underused tool can be termed a tool change policy (TCP) and is illustrated in Section 4. In an intelligent tool condition monitoring (TCM) system, the definition and implementation of a TCP should be carried out autonomously and independently.

TCM is one of the classical problems of manufacturing, and it has been extensively studied in the last four decades [7,8]. However, three barriers have been identified by the presented study that challenge the deployment of existing solutions in machine shops around the world. First, the inflexibility of the systems to accommodate different TCPs: tool condition affects different aspects of machining like surface finish [9] and dimensional accuracy [10,11]. For example, a TCP for one tool is when the chatter marks start to appear, while for another tool TCP it is related to a burr on the edge. The existing studies fail to provide flexibility to accommodate different TCPs.

Quantification of tool wear ignores the concept of a TCP and diverts the attention to the quantification of wear on inserts, and this is the second identified challenge for deployment of current TCM systems. The studies try to quantify the wear on inserts in terms of millimeters of flank wear [12–15]. This quantification provides no information about the usability of the tool. In machine shops around the world, quality management is not seen as a process that directly adds value to the component, and from an economic point this process must be limited to what is absolutely necessary [16]; that is why the manufacturers are interested to know whether components meet the design requirements (a conforming part) or do not meet design requirements (a non-conforming part). One of the examples for this qualitative approach is GO (conforming part)/NO GO (non-conforming part) gauges [17], which are discussed in Section 3. Therefore, the central objective of TCM must also be qualitative so that it recognizes the GO quality tool (tool that produces conforming part) and the NO GO quality tool (tool that produces non-conforming part).

The final barrier identified is the large amount of data and time required to collect and train these systems, which is the most significant barrier in the accommodation of different TCPs. The models have to be retrained for different quality requirements that require changing the parameters learned by the predictive systems. For example, Wu et al. [15] used 5880 images to train a model for the detection of different wear patterns. Considering four cutting edges per insert, the model used 1470 inserts for training. Collecting these extensive data for every machine and TCP is infeasible considering the hundreds of different quality requirements in machine shops around the world.

The proposed system is an integrated solution to the three barriers mentioned above. The system relies on monitoring the wear of cutting tools and classifies the tools as GO/NO GO tools that help the machine operator make the decision on whether the tool can be used for the next machining cycle. The system uses state-of-the-art tool wear classification in the form of a convolution neural network (CNN) and principles of transfer learning (TL). These concepts are discussed in Section 3. The novelty of the system is its ability to correlate the tool condition with machining quality and the accommodation of different quality requirements using a TCP with the requirement of fewer data to achieve the accommodation.

The rest of the research paper is structured as follows. In Section 2, the relevant studies are discussed. Section 3 explains the methodology used in the system, which can be divided into training, offline state, and online state. In Section 4, the case study and the implementation of the proposed system are discussed and the evaluation of the proposed system is performed in Section 5, followed by the suggested future direction of tool condition monitoring. In the final section, the conclusions drawn from the study are discussed.

2. Literature Review

Tool condition monitoring methods are classified into direct and indirect methods [18]; direct methods mostly involve the use of computer vision [14,15], radiation [19], and electrical resistance [20], whereas indirect methods involve online monitoring methods that use vibration [21–23], force [22], and temperature and sound [24] signals. Indirect methods are less complicated and can be implemented straightforwardly and monitored in real-time [10], but they are prone to making noise and are less accurate than the direct methods [25]. Real-time monitoring is also not a crippling disadvantage for direct systems as there is enough time in between machining operation and cycles [26] to get the required data without disturbing the sequence of operations of a machine shop. In addition, the unidirectional execution of existing G-code-based systems does not allow for real-time changes in the machining parameters [27–29]; therefore, there is no way to integrate the response generated by indirect systems in real-time. Considering direct methods are more accurate systems, the study adopts the direct monitoring methodology.

Vision-based systems are the most popular systems when it comes to direct tool condition monitoring. Vision systems have also improved in recent years and are being used in different facets of machining like collision avoidance [2,30], which also demonstrates the ability of vision systems to detect changes while maintaining distance from the cutting process. Computer vision systems are used to monitor changes in the wear morphologies of an insert. Wear morphology classification has been the subject of many studies in past years; Lanzetta [31] employed vision systems for wear morphology classification using quantitative definitions of different wear patterns. The conventional tool condition monitoring used in machine shops involves the quantitative approach. For autonomous systems, the quantitative approach proves difficult for implementation considering the variety of qualitative parameters that need to be hardcoded into the system to identify a variety of wear morphologies. The hard coding of parameters is also computationally expensive; thus, there is a need for a system that identifies the features of different types of wear. The need for identification of different morphologies is satisfied to an extent using CNN by Wu et al. [15]; this study has inspired the base wear classification model discussed in Section 3.1.

Autonomously detecting damage to the inserts before they are used in machining is one of the essential requirements for making tool monitoring autonomous. Fernandez-Robles et al. [32] developed a vision-based system to detect broken inserts in milling cutters automatically. Sun et al. [14] used image processing and image segmentation techniques to develop a system that could identify built-up edges, fractures, and other insert deformations. These studies used image processing techniques, which require human intervention to develop feature descriptions; this limits the independent implementation of these systems for a variety of wear morphologies. As opposed to image processing techniques, the CNN approach learns to identify the region of interest (ROI) and features descriptions to identify different wear morphologies, and this eliminates the need for the human feature descriptions step needed in other techniques [33]. Considering the utility of autonomous feature extraction, the proposed study uses the CNN approach for tool condition monitoring.

Even though tool condition monitoring is one of the classical problems, there are fewer publications in the context of the correlation of tool condition with the quality of the component. Jain and Lad [34] developed a system that correlates tool condition and production quality. The study also developed a multi-level categorization of the wear using a support vector machine methodology. Jain and Lad [35] explored the relationship between surface finish and tool wear and found the Pearson correlation coefficient between surface finish and tool condition to be significant to establish a strong correlation. The study used a random forest-based fault estimation model to determine the relation between surface finish and tool condition. Grzenda and Bustillo [36], developed a semi-supervised model to predict the surface finish using vibration signals; Fourier transformation was used to transform signals to frequency space, and only the relevant frequency ranges were considered for the study. Wu et al. [15] developed a two-stage system that aimed to determine the type of wear in the first stage and tried to quantify the wear of the insert in a milling cutter. The system used CNN to determine the type of wear, and the wear value was obtained using the relation between image pixel value and actual value and width of the minimum circumscribed rectangle. Dutta et al. [37] used surface texture descriptions to determine tool life using the grey level co-occurrence matrix; the images of the resulting surface finish were captured, and the tool wear was measured using a microscope. García-Ordás et al. [25] used a computer vision system to determine the usefulness of milling cutters. The system used a support vector machine methodology to classify the wear patterns. The system identified the state of the tool with about 90 percent accuracy.

The studies mentioned above correlate tool condition with the specific quality and design requirements like surface finish. As discussed in Section 1, there are a variety of quality and design requirements that are defined by TCP. These different TCPs form the ultimate definition of TCM; considering this, a TCM that is flexible enough to accommodate different TCPs is the need of the hour. Most of the studies are also limited by the materials and tool geometries they have used, and changing any one of the factors means the findings of the studies cannot be used. For the TCM to be autonomous and independent, it must be capable of working with different materials, tool geometries, and tool coating grades. The requirements of flexibility to work with different TCPs and the ability to generalize the system for different working materials, tool geometries, and tool coating grades, form the basis for the development of the proposed system.

3. Qualitative Tool Condition Monitoring System

The system is developed to operate in three stages, as shown in Figure 1. The training of the base model is where the architecture of the base model and the central intelligence of the system are developed; this training is done remotely. The architecture and training parameters of the base model are discussed in Section 3.1. The offline state of the system is operational in the machine shops when the production lines are set up. In this state, the system is receiving training to identify the TCP. The knowledge from the base wear classification model is used to expedite this training process using the TL technique discussed in Section 3.2. The output of the system is inspired by the GO/NO GO gauges. The goal of the GO gauge is to accept as many good parts as possible that satisfy the material condition specification, and NO GO gauges are designed to reject all the parts that violate the material condition specification [17]. The GO/NO GO gauge in this system is envisioned as an implementation of a TCP. When a tool of GO quality is detected, the tool is accepted and used for production. When a NO GO quality tool is detected, the operator is asked to change the tool before resuming the production. The GO/NO GO arrangement allows for the flexibility to adapt the system for different TCPs. In the online state of the system discussed in Section 3.3, the system is executing the TCP autonomously after every machining cycle, making sure the tools are in GO condition before they are used, and in this way provides a proactive approach to TCM.

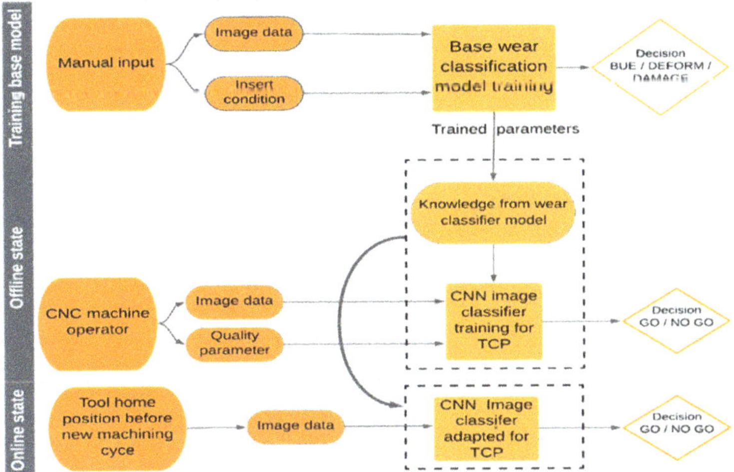

Figure 1. Overview of the system architecture.

3.1. The Base Wear Classification Model

A CNN is one of the most promising approaches to image processing and pattern recognition [33]. CNN layers are part of the architecture; it is standard practice to use convolution layers at the start of the model to develop feature descriptions of the images. These layers are good at narrowing down the ROI and require less computational memory when compared to conventional models. These are the reasons they have seen a wide range of applications in a variety of areas, from hand gesture recognition [38] to disease recognition in plants [39]. One of the other advantages of the convolution layers is their ability to extract features autonomously. Some of these transformations are shown in Figure 2. In Figure 2b, each row is the output of convolution or max pooling layers. It can be seen in the successive layers. The layer transformation and filtering further define the description of the wear features. This step in image processing techniques is done manually, which is the disadvantage of image processing techniques.

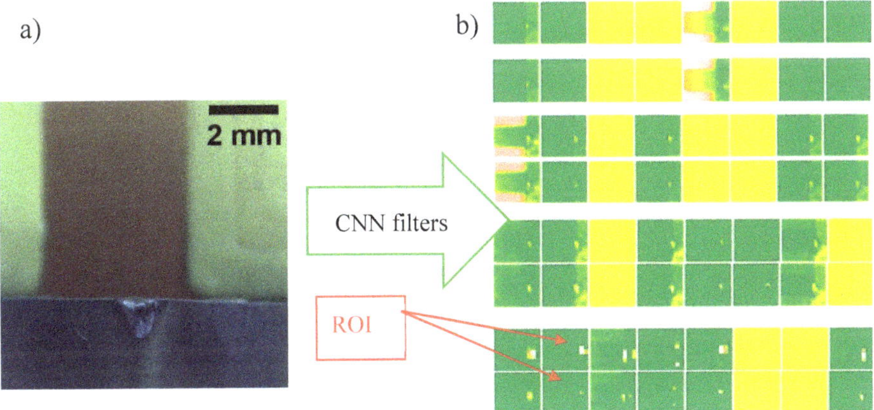

Figure 2. Convolution layer image transformation: (**a**) original image of wear; (**b**) convolution transformation for region of interest (ROI) and feature extraction.

The base model architecture used by the system is shown in Figure 3. The input to this architecture is a $200 \times 200 \times 3$ Red Green Blue image (RGB). The convolution layers have sparse interaction with the input of the previous layer [40]. For the convolution layer, a 3×3 kernel is used with 32 filters in the first and second convolution layers. For the last two convolution layers, 64 filters are used with a 3×3 kernel. A kernel can be imagined as a 3×3 window sliding over 200×200 in the step of a one-pixel slide. This concept helps in the detection of small meaningful features and also reduces the parameters to be stored and computed [40]. The output of the convolution layer is then fed to the pooling layer, in the case of the developed model, it is the max pooling layer, where the kernel reports the maximum value of the kernel size input. This layer helps in making the model more robust in response to small translations to the inputs [40]. This is summarized in Equation (1), where A is the corresponding pixel value in row i and column j, and this equation is valid for the 2×2 kernel used for max pooling layers.

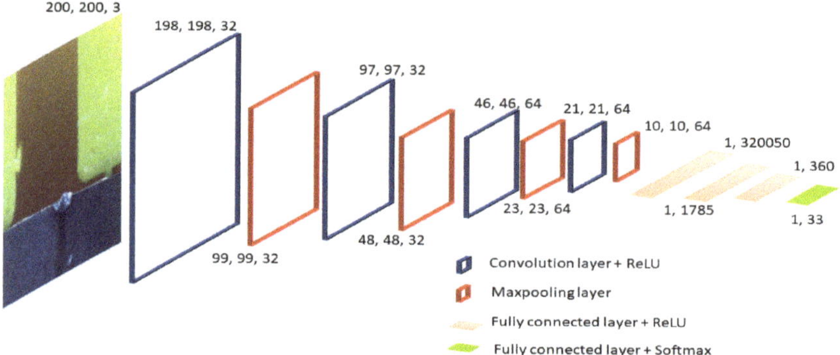

Figure 3. Base convolution neural network (CNN) architecture for insert wear classification.

Max pooling layer

$$O_{i,j} = \max \{A_{i,j}, A_{i+1,j}, A_{i,j+1}, A_{i+1,j+1}\} \tag{1}$$

The output of the last max pooling layer is then flattened to a $1 \times n$ vector, which forms the input to the fully connected layers (FCLs) for further processing, where n is the number of inputs to the neural network. The number of inputs n also determines the width of the FCLs of the network.

FCLs are the basic type of neural network where each input interacts with each output of the previous layer [40], the different layers in the network are modeled as different functions, which are the function of the previous layer. In the proposed base model shown in Figure 3, we have four FCL layers, which can be written as $f^{(1)}$, $f^{(2)}$, $f^{(3)}$, and $f^{(4)}$. Using the chain concept we can rewrite these functions as $f(x_i) = f^{(4)}(f^{(3)}(f^{(2)}(f^{(1)}(x_n))))$ [40], where x_n is the data from the convolution layers. The objective of the neural network is to best estimate $f(x_n; \theta)$ to function $f^*(x_n)$, where $f^*(x_n)$ is the ideal (real-world relation) function that maps the inputs from the convolution layer to their classes of wear and θ is a free parameter adjusted to optimize the best estimation of an ideal function [40].

The architecture in Figure 3 uses a rectified linear unit (ReLU) activation function in the intermediate layers, which is a common practice for CNNs to improve the training speed [39]. The ReLU returns zero for half of its domain and is the input for the other half of the domain that is zero for inactive nodes and is the node output for active nodes, which helps make the gradients of the loss function large and constant [40]. The ReLU is used in all the layers except the output layer in the proposed model shown in Figure 3. The ReLU is summarized in Equation (2), where z is the output of the node.

$$ReLU(z) = \max\{0, z\} \tag{2}$$

The softmax activation function was used in the last layer of the base model architecture, which is also common in multiclass classification CNNs [39]. Softmax activation usually used in the output layers of the neural networks gives the probability distribution over n possible values. It ensures that the prediction of z belonging to a class for n different classes is between 0 and 1, and the sum of probabilities is equal to 1 [40]. This is summarized in Equation (3), where z_p is the output of the node for the p class.

$$\text{softmax}\left(z_p\right) = \frac{e^{z_p}}{\sum_{c=1}^{3} e^{z_c}} \tag{3}$$

The loss of a model can be defined as a function that quantifies the performance of the system [33]; the study uses categorical cross-entropy as the loss function. ADAM, which is a stochastic optimizer that is computationally efficient and combines the advantages of RMSProp and AdaGrad [41], was used to optimize the weights of the base model. This facilitated faster convergence to an optimal solution [40]. The parameters used in ADAM were learning rate = 0.001, $beta^1$ = 0.9, and $beta^2$ = 0.999. The algorithm for ADAM implementation can be found in [41].

3.2. CNN Image Classifier Trained for TCP

The base model developed and discussed in Section 3.1 forms the central intelligence for the TCM. The base model helps narrow down the ROI and extract useful features and descriptions of the tool, as shown in Figure 2. This intelligence is developed in the base model and rolled out as a trained network. The offline stage of the system is in the machine shops, where the model has to be repurposed to identify and implement different TCPs. Considering that there are thousands of different TCP unique to each machine shop, retraining a complete network presents a significant data and training time challenge. TL is one of the lifelines to overcome this data and training time challenge.

Given the importance of TL, we now adapt the definitions of TL in [42] for our application. In the proposed system, the knowledge developed during the training of the wear classification model to identify what type of wear pattern or damage the cutting tool has is optimized using TL to differentiate between a good tool that produces conforming parts and a bad tool that produces non-conforming parts. Every classification model has a domain D, which forms the pool for data extraction and a task which, in the case of this study, is classification. Pan and Yang [42] define domain D as consisting of two components: a feature space X and a marginal probability P(X). Task T also consists of two-component Y labels and a predictive function f(.); since neural networks have a large number of trainable parameters they can choose from different functions that best predict the task, which in the case of our study is image classification. Therefore $D = \{X, P(X)\}$ and $T = \{Y, f(.)\}$, and considering these definitions we can define source domain and target domain. The source domain is images captured from cutting inserts used in machining (D_S), and the task is to identify wear type classification (T_S). Similarly, the target domain is images of inserts used in production (D_T), and the task is the quality classification (T_T).

The images for the base model are drawn for the inserts used in production. Similarly, images used for the target model are also drawn from inserts used in production. Therefore, the methodology is built around the assumption that the images for the source and target model have a similar domain, which is reasonable considering that the images used to train base model wear morphology classification are also used in production in a machine shop. Given the similarity of domains, $X_S = X_S$, $P_S(X) = P_T(X)$, and $D_S = D_T$, that is, the feature space and the marginal probability of data distribution for both models are the same.

The tasks of the source and target models are different as the labels are different, therefore $T_S \neq T_T$ as $Y_S \neq Y_T$, as given by Equations (4) and (5). But the predictive function could be similar or different since the neural networks are black-box models. There is no way to know if the same or a different function was used for source and target tasks.

$$Y_S = \begin{cases} 0 & \text{if the insert is damaged} \\ 1 & \text{if the insert has deformation} \\ 2 & \text{if the insert has normal wear} \end{cases} \tag{4}$$

$$Y_T = \begin{cases} 0 & \text{if the component is conforming (GO)} \\ 1 & \text{if the component is non} - \text{conforming (NO GO)} \end{cases} \tag{5}$$

The target task is tied up to the traditional concepts of GO/NO GO gauges discussed at the start of Section 3. GO/NO GO gauges are one of the most popular gauges to evaluate the material conditions in holes and shafts. The tool condition monitoring system developed extends and generalizes this definition of GO/NO GO gauges to other quality requirements. In the target task, the model is retrained to identify a GO part producing tool and NO GO part producing tool. This concept makes the proposed methodology qualitative and gives the model the flexibility to adapt its knowledge across different TCPs. The offline state requires an expert to generate the GO/NO GO labels for the training and adaptation of the task to the TCP.

3.3. CNN Image Classifier Adapted for TCP

The online state of the system works seamlessly without the need for human intervention to identify the tools that produce a NO GO part. The system takes a picture before the machining starts and, based on the training during the offline state, classifies the tool as a useable or unusable tool. There are many tools on the machine, and the quality demand from each tool is different. Therefore, the offline part of the system where the tool condition is associated with GO/NO GO quality of operation has to be performed on each tool during the production setup. This allows the system to run without quality inspection in the online state.

4. Experimental Setup

The images of the CNMG 120408/12, TNMG 160408/12, and VNMG 160408 turning inserts used in the turning application are captured using the DFK 33GP006 GigE color camera with TCL 3520 5MP lens. Initially, the top, side, and front views are considered for the classification. A processor with Intel i7 and 16 GB RAM was used to develop the classification model, and the models were implemented using R computer language with the Keras package using a TensorFlow backend. Examples of these pictures can be seen in Table 1. Table 1 shows that the top and side views do not clearly show the type of wear, but the wear is easily distinguishable in the front view images; therefore, only front view images were considered for the classification model. The setup for capturing the images of different views can be seen in Figure 4. The images are captured in standard room lighting without any dedicated light source. As we can see from Figure 2, the background of the insert has no impact on the feature extraction process.

Table 1. Wear patterns and different views.

Wear Pattern	Top View	Side View	Front View
Damaged			
Deformation			
Abrasive			

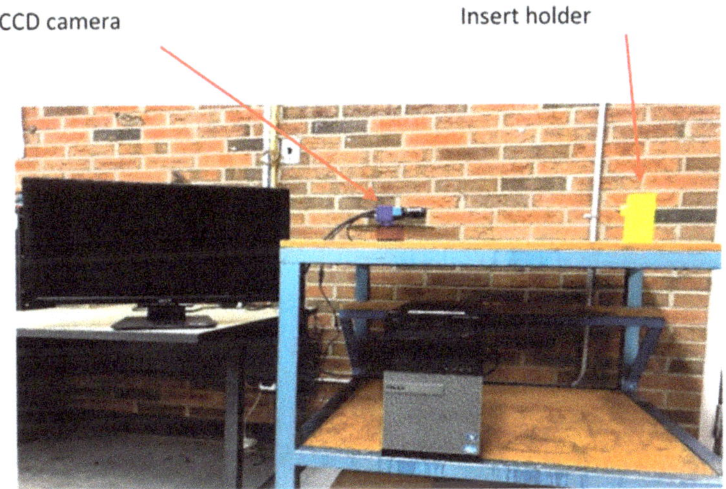

Figure 4. Image capturing station.

The process started with collecting the front-view images of the inserts for three different categories: 79 images of damaged inserts, 121 images of deformed inserts, and 128 images of abrasive wear inserts were captured. All the pictures were then resized to 200 (width) × 200 (height).

The model must be made robust against variation and transformation. One of the ways to do this is to use data augmentation, where the data are subjected to various transformations such as rotation, flipping, and shearing the images; this helps in improving the generalization error as the model is trained to be invariant to these transformations [40]. The training dataset was subjected to data augmentation, with an allowed rotation range of 10 degrees, width shift range of 20 percent, height shift range of 10 percent, and zooming range of 20 percent. The parameters for augmentation were chosen carefully so as not to alter the wear description of the images, but to accommodate for poor quality images that can be seen when the system is deployed in machine shops. An example of this data augmentation can be seen in Figure 5. It should be noted that the validation images were not subjected to data augmentation.

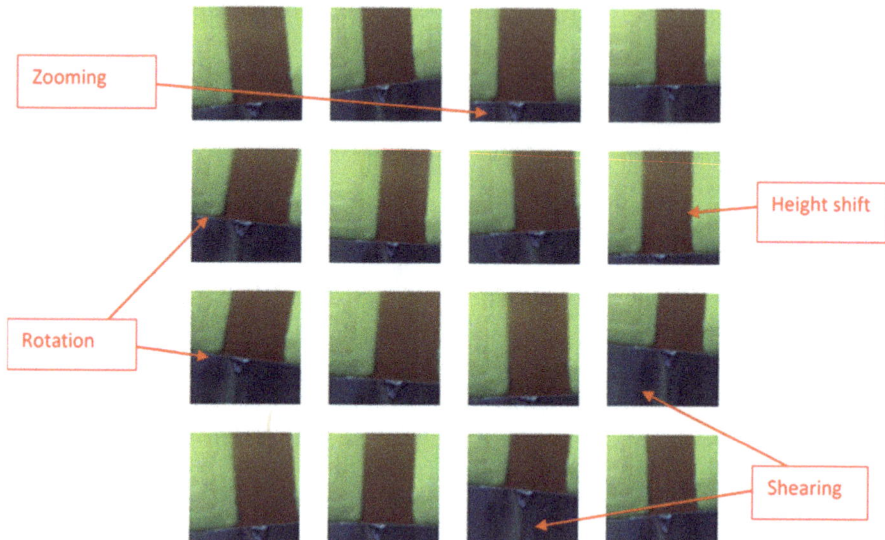

Figure 5. Examples of rotation, shearing, height shift, and zooming data augmentation applied to training images.

The augmented data were then fed to the neural networks. The images were first subjected to the image transformations of the convolution and max pooling layers, where the ROI was identified, and useful features were extracted from the images autonomously. The data in the final layers formed the last max pooling layer and formed the input to the densely connected layers. The images were converted to pixel data, and each pixel formed the input to the first FCL; the data were mapped to the labels of the pictures, and the output of the FCL was the prediction of wear type. The base model can now identify the nuanced differences in the cutting insert by identifying if the insert has deformation, normal wear, or damage. The knowledge developed in the base model can now be used to identify the change in quality by retaining the model.

For the target model, the images of inserts are classified into GO/NO GO categories; for this part of the study, only CNMG 120408 turning inserts are used; one of the examples for GO/NO GO can be seen in Table 2. Table 2 also gives an example for different TCPs accessible in machine shops. The objective of the case study with respect to the target model is to prove that the model identifies the nuanced differences in changes in wear morphology and predicts the consequence of using a tool. That is, if the tool produces a GO quality part or NO GO quality part while using lesser images and training time and iterations so that the TCP deployment is fast-tracked.

For this part of the study, inserts that are relatively new and have typical wear patterns are manually classified as GO category inserts, and the inserts that have higher wear levels, as shown in the NO GO part of Table 2, are classified as NO GO category inserts. These images are used to train the target model, and the training images are subjected to similar data augmentation to that shown in Figure 5. The architecture of the target model is shown in Table 3. The parameters learned by the base model are frozen, and only 382 parameters of layers 12 and 13 are optimized for the target model.

Table 2. GO, NO GO, and examples of different tool change policies used in machine shops around the world.

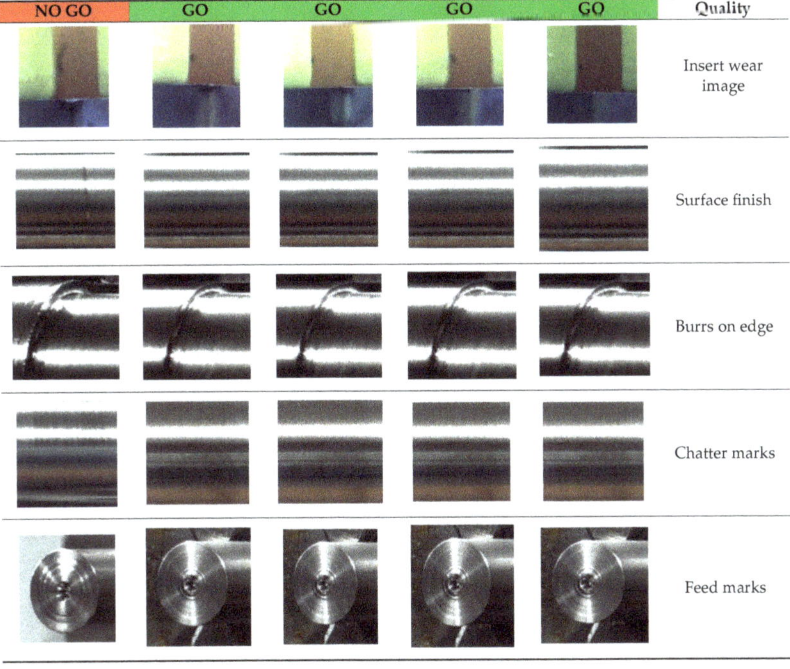

NO GO	GO	GO	GO	GO	Quality
					Insert wear image
					Surface finish
					Burrs on edge
					Chatter marks
					Feed marks

Table 3. Target model architecture.

	Layer Type	Input Shape	Output Shape	Filter Size	Trainable Parameters
0	Input layer	200,200,3	200,200,3	0	
1	Convolution layer	200,200,3	198,198,32	3,3,32	0
2	Max pooling layer	198,198,32	99,99,32	2,2,32	0
3	Convolution layer	99,99,32	97,97,32	3,3,32	0
4	Max pooling layer	97,97,32	48,48,32	2,2,32	0
5	Convolution layer	48,48,32	46,46,64	3,3,64	0
6	Max pooling layer	46,46,64	23,23,64	2,2,64	0
7	Convolution layer	23,23,64	21,21,64	3,3,64	0
8	Max pooling layer	21,21,64	10,10,64	2,2,64	0
9	flatten (Flatten)	10,10,64	6400,1	0	0
10	Dense layer	6400,1	50,1	0	0
11	Dense layer	50,1	35,1	0	0
12	Dense layer	35,1	10,1	0	360
13	Dense layer	10,1	2,1	0	22

5. Results and Discussion

The training of the base model was carried out using 223 images, and 105 images were split from the original dataset for validation; the validation data set consists of approximately 33 percent from each of the damaged, deformation, and abrasive wear categories. Figure 6a gives the accuracy for base model training runs. The validation accuracy stabilized around the 200th epoch, and the validation accuracy is 83.75 percent. Figure 6b presents the loss of over 250 epochs, and the loss is the indication of the magnitude of deviation between prediction and the actual value.

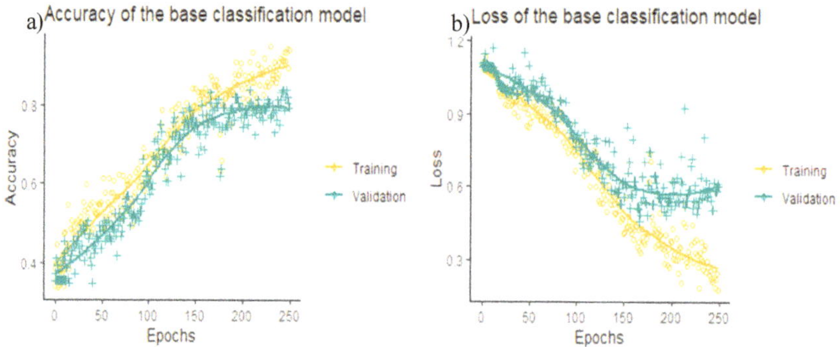

Figure 6. (a) Accuracy of the base model. (b) Loss convergence graph of the base model.

For the second part, the objective is to demonstrate the capability of the system to adapt to the new task of TCP deployment using fewer images and shorter training time. For this, the data are partitioned into three sections: training, validation, and test dataset; various training runs are carried out using a different number of images. The summary of the number of images used for each run is shown in Table 4. All the images in the three sections are different and are not repeated. The images in the test data set can be seen in Figure 7; the GO category images have no wear or have typical wear pattern; these tools produce conforming parts, and the NO GO category have visible wear on the edges; these tools produce non-conforming parts.

Table 4. Training, validation, and test data split for different runs.

RUN	Training Pictures	Validation Pictures	Test Pictures
1	13	10	20
2	18	12	20
3	23	16	20
4	27	18	20
5	37	26	20

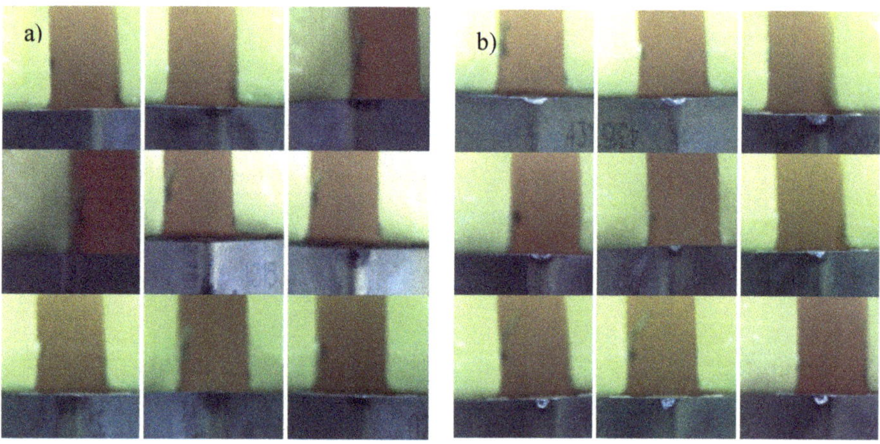

Figure 7. Test image data: (a) GO images; (b) NO GO images.

Figure 8 shows the accuracy and loss values for different runs. It can be seen from Figure 8b that the loss value plateaued around the 5th epoch in most of the runs, signifying that the optimization of the parameters requires fewer iterations, which enables the system to accommodate a variety of

TCPs and with fewer training requirements. Figure 8c and d show the accuracy and loss of the trained models on the test dataset. It can be seen that the accuracy of the models increases with the number of images used for training the model. Run 4 had a smaller loss value when compared to run 5 but had lower accuracy; this can be attributed to overfitting of data which led to misclassification of the images in the NO GO category test data as GO category images. Run 5 had the best results in terms of accuracy on test data, where 37 images were used in training the model. The accuracy of the test data for run 5 was 85 percent; the confusion matrix for run 5 can be seen in Table 5. The model predicted all NO GO label images correctly and predicted 3 images of GO labels incorrectly.

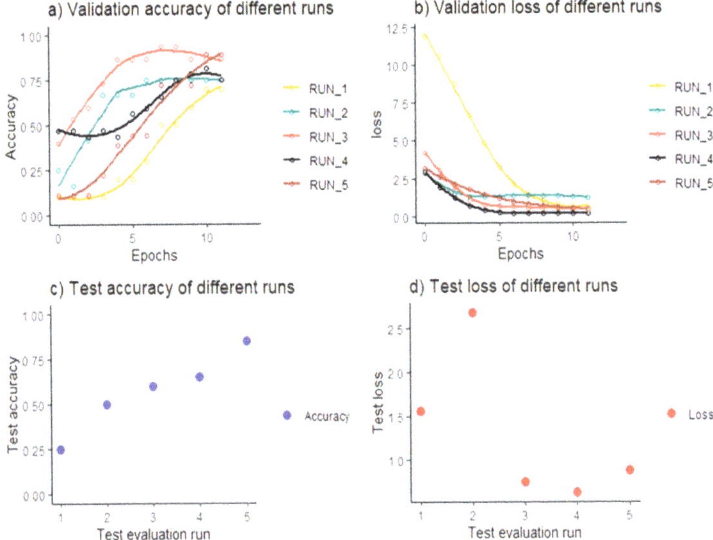

Figure 8. Results for target model training run on validation and test datasets.

Table 5. Confusion matrix for run 5.

Actual \ Prediction	GO	NO GO
GO	7	0
NO GO	3	10

The final part of the study is the deployment of the system using a graphical user interface (GUI). Figure 9 gives a view of the GUI; the output of the GUI is feedback to the operator. The feedback is NO GO for tools that the target model predicts will produce a non-conforming part, and GO for tools that the model predicts will produce a conforming part. The machine operator is encouraged to replace the tool when the GUI displays NO GO. The prediction is generated within 5 s, facilitating the mass production without interruptions from quality inspections.

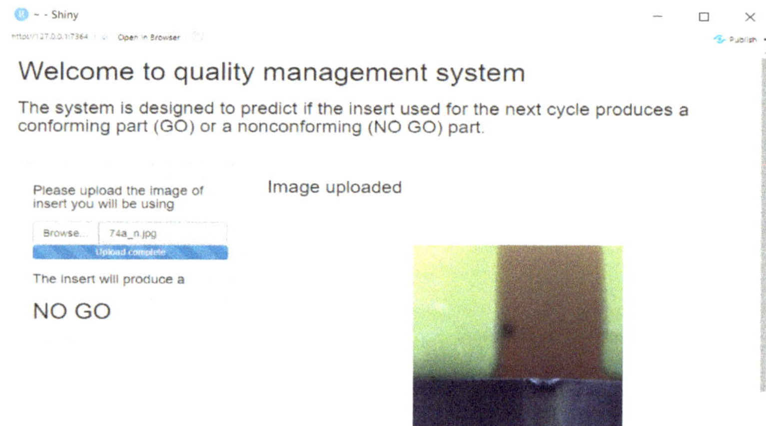

Figure 9. Graphical user interface (GUI) for quality management system.

The proposed system directs the TCM from a quantitative to a qualitative approach using a TCP. This is the reason the study does not consider quantification of the wear using any measuring system, and this makes the system more flexible to accommodate different quality requirements seen in machine shops around the world. The system uses the feature extraction capabilities of CNN and the ability of these models to learn new features using TL. Future development in the proposed system has three fronts. First, the development of the camera system to be integrated into the machine: the image acquisition in the study was independent of the machine, but the system has demonstrated that the images acquired at a reasonable distance away from the cutting operation can be used and classified by the system. The system uses standard room lighting; there are systems discussed in studies done by Sun et al. [14], which are capable of generating the required light intensity and protecting the camera from cutting oil. Second, improvement of the intelligence and accuracy of the system by integrating more diverse wear images into the base model: since the base model is the central nervous system and target models have similar data distribution, the accuracy of the GO/NO GO model can be improved with fewer iteration requirements, and with a more robust base model. Finally, the proposed system is designed for the turning process: considering that the wear mechanisms in milling are different, there is a need to develop a similar system for milling applications. A similar neural network can be trained with milling insert data to repurpose the framework for a milling application. The CNN architecture is a standard approach when it comes to image recognition and identification-related neural network architectures. The area of deep learning also continues to evolve; therefore, there is a need to keep an eye out for new techniques that can improve the training time and accuracy of the system.

6. Conclusions

The proposed study redirects tool condition monitoring from tool wear quantification to the objective of providing a more proactive qualitative approach to quality management that saves the resources used in the production of non-conforming parts. A tool change policy is employed to detect these changes in quality indicators and change the tools when they occur. Given the different quality requirements, there are a variety of tool change policies. Therefore, tool condition monitoring that is flexible enough to adapt to different tool change policies is developed. The developed systems can adapt to a new tool change policy requiring as few as 37 images and 12 training iterations using concepts of transfer learning and convolution neural networks. The study also developed a qualitative approach to tool condition monitoring, since the issue that is more important than quantifying the wear on a cutting tool is whether the tool produces a conforming or a non-conforming part. This is captured in the system by extending the concept of GO/NO GO gauges to different quality requirements through tool conditions. The tools that are predicted to produce a conforming part are categorized as GO tools, while non-conforming part producing tools are classified as NO GO tools. The developed system can identify the GO/NO GO quality tool with 85 percent accuracy. Lastly, a graphical user interface is also developed to give feedback to machine operators about the usability of the tools. The system is designed to operate in between machining cycles, checking the usability of every tool before they are used. The system only takes a few seconds to determine the usability of the tools. This concept helps in making the proposed methodology an in-process tool condition monitoring system.

Author Contributions: Conceptualization, H.M.; methodology, H.M. and M.A.S.; software, H.M.; validation, H.M. and M.A.S.; formal analysis, H.M.; investigation, H.M.; resources, R.A.; data curation, H.M.; writing—original draft preparation, H.M.; writing—review and editing, R.A.; visualization, H.M.; supervision, R.A.; project administration, R.A.; funding acquisition, R.A. All authors have read and agreed to the published version of the manuscript.

Funding: We express our gratitude to the Minister of Economic Development, Trade, and Tourism for funding this project through Major Innovation Funds. The authors also would like to acknowledge the NSERC (Grant Nos. NSERC RGPIN-2017-04516 and NSERC CRDPJ 537378-18) for further funding this project.

Acknowledgments: We express our gratitude to the Minister of Economic Development, Trade, and Tourism for funding this project through Major Innovation Funds. The authors also would like to acknowledge the NSERC (Grant Nos. NSERC RGPIN-2017-04516 and NSERC CRDPJ 537378-18) for further funding this project.

Conflicts of Interest: The authors declare no conflict of interest.

References

1. Albers, A.; Gladysz, B.; Pinner, T.; Butenko, V.; Stürmlinger, T. Procedure for Defining the System of Objectives in the Initial Phase of an Industry 4.0 Project Focusing on Intelligent Quality Control Systems. *Procedia CIRP* **2016**, *52*, 262–267. [CrossRef]
2. Ahmad, R.; Tichadou, S.; Hascoet, J.Y. Generation of safe and intelligent tool-paths for multi-axis machine-tools in a dynamic 2D virtual environment. *Int. J. Comput. Integr. Manuf.* **2016**, *29*, 982–995. [CrossRef]
3. Lee, J.; Bagheri, B.; Kao, H.A. A Cyber-Physical Systems architecture for Industry 4.0-based manufacturing systems. *Manuf. Lett.* **2015**, *3*, 18–23. [CrossRef]
4. Wang, G.F.; Yang, Y.W.; Zhang, Y.C.; Xie, Q.L. Vibration sensor based tool condition monitoring using ν support vector machine and locality preserving projection. *Sens. Actuators A Phys.* **2014**, *209*, 24–32. [CrossRef]
5. Khorasani, A.M.; Gibson, I.; Goldberg, M.; Doeven, E.H.; Littlefair, G. Investigation on the effect of cutting fluid pressure on surface quality measurement in high speed thread milling of brass alloy (C3600) and aluminium alloy (5083). *Meas. J. Int. Meas. Confed.* **2016**, *82*, 55–63. [CrossRef]
6. Pratama, M.; Dimla, E.; Lai, C.Y.; Lughofer, E. Metacognitive learning approach for online tool condition monitoring. *J. Intell. Manuf.* **2019**, *30*, 1717–1737. [CrossRef]
7. Ghani, J.A.; Rizal, M.; Nuawi, M.Z.; Ghazali, M.J.; Haron, C.H.C. Monitoring online cutting tool wear using low-cost technique and user-friendly GUI. *Wear* **2011**, *271*, 2619–2624. [CrossRef]
8. Jinpeng, S.; Shaowei, L.; Ahmad, R.; Jiaojiao, G.; Ming, L. Tribological behaviour of TiB2-HfC ceramic tool material under dry sliding condition. *Ceram. Int.* **2020**, *46*, 20320–20327. [CrossRef]
9. Jinpeng, S.; Jiaojiao, G.; Ahmad, R.; Ming, L. Cutting performances of TiCN–HfC and TiCN–HfC–WC ceramic tools in dry turning hardened AISI H13. *Adv. Appl. Ceram.* **2020**. [CrossRef]
10. Teti, R.; Jemielniak, K.; O'Donnell, G.; Dornfeld, D. Advanced monitoring of machining operations. *CIRP Ann. Manuf. Technol.* **2010**, *59*, 717–739. [CrossRef]
11. Kilundu, B.; Dehombreux, P.; Chiementin, X. Tool wear monitoring by machine learning techniques and singular spectrum analysis. *Mech. Syst. Signal Process.* **2011**, *25*, 400–415. [CrossRef]
12. Mikołajczyk, T.; Nowicki, K.; Bustillo, A.; Pimenov, D.Y. Predicting tool life in turning operations using neural networks and image processing. *Mech. Syst. Signal Process.* **2018**, *104*, 503–513. [CrossRef]
13. D'Addona, D.M.; Teti, R. Image data processing via neural networks for tool wear prediction. *Procedia CIRP* **2013**, *12*, 252–257. [CrossRef]
14. Sun, W.H.; Yeh, S.S. Using the machine vision method to develop an on-machine insert condition monitoring system for computer numerical control turning machine tools. *Materials* **2018**, *11*, 1977. [CrossRef]

15. Wu, X.; Liu, Y.; Zhou, X.; Mou, A. Automatic identification of tool wear based on convolutional neural network in face milling process. *Sensors* **2019**, *19*, 3817. [CrossRef]

16. Orzan, I.A.; Buzatu, C. *CONAT 2016 International Congress of Automotive and Transport Engineering*; Springer: Berlin/Heidelberg, Germany, 2017. [CrossRef]

17. Meadows, J.D. Measurement of Geometric Tolerances in Manufacturing. In *Manufacturing Engineering and Materials Processing*; CRC Press: New York, NY, USA, 1998; pp. 1–5. ISBN 9780429165504.

18. Azouzi, R.; Guillot, M. On-line prediction of surface finish and dimensional deviation in turning using neural network based sensor fusion. *Int. J. Mach. Tools Manuf.* **1997**, *37*, 1201–1217. [CrossRef]

19. Jetley, S.K. Applications of surface activation in metal cutting. In Proceedings of the Twenty-Fifth International Machine Tool Design and Research Conference, Birmingham, UK, 22–24 April 1985.

20. Wilkinson, A.J. Constriction-resistance concept applied to wear measurement of metal-cutting tools. *Proc. Inst. Electr. Eng.* **1971**, *118*, 381. [CrossRef]

21. Wu, T.Y.; Lei, K.W. Prediction of surface roughness in milling process using vibration signal analysis and artificial neural network. *Int. J. Adv. Manuf. Technol.* **2019**, *102*, 305–314. [CrossRef]

22. Kene, A.P.; Choudhury, S.K. Analytical modeling of tool health monitoring system using multiple sensor data fusion approach in hard machining. *Meas. J. Int. Meas. Confed.* **2019**, *145*, 118–129. [CrossRef]

23. Khorasani, A.M.; Yazdi, M.R.S. Development of a dynamic surface roughness monitoring system based on artificial neural networks (ANN) in milling operation. *Int. J. Adv. Manuf. Technol.* **2017**, *93*, 141–151. [CrossRef]

24. da Silva, R.H.; da Silva, M.B.; Hassui, A. A probabilistic neural network applied in monitoring tool wear in the end milling operation via acoustic emission and cutting power signals. *Mach. Sci. Technol.* **2016**, *20*, 386–405. [CrossRef]

25. García-Ordás, M.T.; Alegre-Gutiérrez, E.; Alaiz-Rodríguez, R.; González-Castro, V. Tool wear monitoring using an online, automatic and low cost system based on local texture. *Mech. Syst. Signal Process.* **2018**, *112*, 98–112. [CrossRef]

26. Siddhpura, A.; Paurobally, R. A review of flank wear prediction methods for tool condition monitoring in a turning process. *Int. J. Adv. Manuf. Technol.* **2013**, *65*, 371–393. [CrossRef]

27. Ahmad, R.; Tichadou, S.; Hascoet, J.Y. 3D safe and intelligent trajectory generation for multi-axis machine tools using machine vision. *Int. J. Comput. Integr. Manuf.* **2013**, *26*, 365–385. [CrossRef]

28. Ahmad, R.; Tichadou, S. Integration of vision based image processing for multi-axis CNC machine tool safe and efficient trajectory generation and collision avoidance. *J. Mach. Eng.* **2010**, *10*, 53–65.

29. Ahmad, R.; Tichadou, S.; Hascoet, J.Y. A knowledge-based intelligent decision system for production planning. *Int. J. Adv. Manuf. Technol.* **2017**, *89*, 1717–1729. [CrossRef]

30. Ahmad, R.; Tichadou, S.; Hascoet, J.Y. New computer vision based Snakes and Ladders algorithm for the safe trajectory of two axis CNC machines. *CAD Comput. Aided Des.* **2012**, *44*, 355–366. [CrossRef]

31. Lanzetta, M. A new flexible high-resolution vision sensor for tool condition monitoring. *J. Mater. Process. Technol.* **2001**, *119*, 73–82. [CrossRef]

32. Fernández-Robles, L.; Azzopardi, G.; Alegre, E.; Petkov, N. Machine-vision-based identification of broken inserts in edge profile milling heads. *Robot. Comput. Integr. Manuf.* **2017**, *44*, 276–283. [CrossRef]

33. Traore, B.B.; Kamsu-Foguem, B.; Tangara, F. Deep convolution neural network for image recognition. *Ecol. Inform.* **2018**, *48*, 257–268. [CrossRef]

34. Jain, A.K.; Lad, B.K. A novel integrated tool condition monitoring system. *J. Intell. Manuf.* **2017**, *30*, 1423–1436. [CrossRef]

35. Jain, A.K.; Lad, B.K. Quality control based tool condition monitoring. In Proceedings of the Annual Conference of the Prognostics and Health Management Society, Coronado, CA, USA, 1 January 2015; pp. 418–427.

36. Grzenda, M.; Bustillo, A. Semi-supervised roughness prediction with partly unlabeled vibration data streams. *J. Intell. Manuf.* **2019**, *30*, 933–945. [CrossRef]

37. Dutta, S.; Datta, A.; Das Chakladar, N.; Pal, S.K.; Mukhopadhyay, S.; Sen, R. Detection of tool condition from the turned surface images using an accurate grey level co-occurrence technique. *Precis. Eng.* **2012**, *36*, 458–466. [CrossRef]

38. Lin, H.I.; Hsu, M.H.; Chen, W.K. Human hand gesture recognition using a convolution neural network. *IEEE Int. Conf. Autom. Sci. Eng.* **2014**, *2014*, 1038–1043. [CrossRef]

39. Sun, X.; Mu, S.; Xu, Y.; Cao, Z.; Su, T. Image Recognition of Tea Leaf Diseases Based on Convolutional Neural Network. In Proceedings of the 2018 International Conference on Security, Pattern Analysis, and Cybernetics (SPAC), Jinan, China, 14–17 December 2018; pp. 304–309. [CrossRef]
40. Goodfellow, I.; Bengio, Y.; Courville, A. *Deep Learning*; MIT Press: Cambridge, MA, USA, 2016.
41. Kingma, D.P.; Ba, J.L. Adam: A method for Stochastic Optimization. *ICLR* **2015**, *2015*, 1–15.
42. Pan, S.J.; Yang, Q. A survey on transfer learning. *IEEE Trans. Knowl. Data Eng.* **2010**, *22*, 1345–1359. [CrossRef]

Publisher's Note: MDPI stays neutral with regard to jurisdictional claims in published maps and institutional affiliations.

Article

Tool Wear Monitoring for Complex Part Milling Based on Deep Learning

Xiaodong Zhang [1], Ce Han [1], Ming Luo [1,2,*] and Dinghua Zhang [1,2]

[1] Key Laboratory of High Performance Manufacturing for Aero Engine, Ministry of Industry and Information Technology, Northwestern Polytechnical University, Xi'an 710072, China; helen203@mail.nwpu.edu.cn (X.Z.); hance@mail.nwpu.edu.cn (C.H.); dhzhang@nwpu.edu.cn (D.Z.)

[2] Engineering Research Center of Advanced Manufacturing Technology for Aero Engine, Ministry of Education, Northwestern Polytechnical University, Xi'an 710072, China

*   Correspondence: luoming@nwpu.edu.cn

Received: 1 September 2020; Accepted: 29 September 2020; Published: 2 October 2020

Abstract: Tool wear monitoring is necessary for cost reduction and productivity improvement in the machining industry. Machine learning has been proven to be an effective means of tool wear monitoring. Feature engineering is the core of the machining learning model. In complex parts milling, cutting conditions are time-varying due to the variable engagement between cutting tool and the complex geometric features of the workpiece. In such cases, the features for accurate tool wear monitoring are tricky to select. Besides, usually few sensors are available in an actual machining situation. This causes a high correlation between the hand-designed features, leading to the low accuracy and weak generalization ability of the machine learning model. This paper presents a tool wear monitoring method for complex part milling based on deep learning. The features are pre-selected based on cutting force model and wavelet packet decomposition. The pre-selected cutting forces, cutting vibration and cutting condition features are input to a deep autoencoder for dimension reduction. Then, a deep multi-layer perceptron is developed to estimate the tool wear. The dataset is obtained with a carefully designed varying cutting depth milling experiment. The proposed method works well, with an error of 8.2% on testing samples, which shows an obvious advantage over the classic machine learning method.

Keywords: tool wear monitoring; milling; complex part; deep learning; autoencoder; deep multi-layer perceptron

---

1. Introduction

Tool wear is a cost driver in machining that affects quality and productivity and adds unscheduled downtime for tool changes and the reworking of damaged parts. Accurate tool wear monitoring is necessary to avoid these unnecessary costs. Tool wear monitoring methods can be categorized into two main groups: direct and indirect methods [1]. Direct methods measure the actual wear value with optical, laser or ultrasonic devices. Although direct methods measure tool wear precisely, they are difficult to implement in real-time machining because, in most cases, the tool wear area is unreachable due to the occlusion of workpiece structure and flood coolant. Indirect methods monitor in-process physical parameters to evaluate wear state, such as force, vibration, acoustic emission, current, power and temperature signals [2]. Nowadays, indirect methods are the most widely used in tool wear monitoring because they are easy to conduct in real time and can obtain acceptable accuracy by using a proper monitoring signals and modeling method.

Indirect tool wear monitoring methods can be divided into two categories: physical-based and data-driven methods. Physical-based methods first develop the hand-designed physical model, i.e., the mathematic relationship between tool wear and measurable physical quantities from the mechanism of machining, and then estimate tool wear value through monitoring signals based on the model. Choudhury and Rath [3] proposed an evaluating approach of milling tool flank wear based on the relationship between average tangential cutting force coefficients and tool wear. Cui [4] discussed the influences of process parameters, tool parameters, and tool wear on tangential cutting force coefficients in his dissertation. A recognition approach for milling tool wear was proposed, based on the relationship between tangential cutting force coefficients and tool wear. This approach was needed to solve the cutting force coefficients through actual cutting forces, and then the tool wear was recognized through the solved tangential cutting force coefficients. The complex calculations meant that the recognition speed was low, and the result was not precise enough. Shao et al. [5] established a cutting power model in face milling, which included the cutting conditions and the tool flank wear, and then proposed a tool wear monitoring approach based on this power model. Hou et al. [6] developed the relationship between flank wear and average milling force based on the stress distribution in the tool wear zone and applied this relationship to estimate the flank wear width in the milling process. Han et al. [7,8] proposed a mechanistic cutting force model considering various wear types for difficult-to-cut material drilling, which can be used for tool wear monitoring and process parameter optimization in the conditions of multiple tool wear existing on different cutting edges.

Due to the complexity of tool wear mechanism and randomness of the machining process, the physical-based monitoring methods are faced with the problem of low accuracy and weak universality [9,10]. In recent years, many data-driven tool wear monitoring methods have been proposed based on machine learning such as fuzzy logic, artificial neural network (ANN), support vector machine (SVM), and Bayesian networks. Yu [11] used logistic regression with penalization and manifold regularization for tool condition monitoring. Kilickap et al. [12] used ANN with the inputs of cutting speed, feed rate and depth of cut for tool wear prediction in the milling of Ti-6242S. Patra et al. [13] used ANN with thrust force signals for tool wear prediction in micro-drilling. Karam et al. [14] built an ANN-based cognitive decision-making system to extract signal features for online tool life prediction. In other studies, the ANNs with different inputs and architectures are employed to predict tool wear [15–17]. Madhusudana et al. [18] used SVM with the features extracted from discrete wavelet transformation in sound signals for tool condition monitoring in face milling. Benkedjouh et al. [19] used support vector regression (SVR) with the features extracted from multi-sensor signals to predict tool life. Zhang and Zhang [20] used a least-square SVM to develop a nonlinear regression model for tool wear prediction in the milling process. Yu et al. [21] proposed a weighted hidden Markov model for tool life prediction. Zhu and Liu [22] proposed a hidden semi-Markov model with dependent durations through cutting force signals for tool wear monitoring. Tobon-Mejia et al. [23] established a dynamic Bayesian network for tool condition monitoring and remaining useful life estimation. Kong et al. [24] proposed a Gaussian process regression model for tool wear prediction. Wu et al. [25] made a comparative study on different machining learning method including ANN, SVR and random forest for tool wear monitoring. Liu et al. [26] built an Elman_Adaboost predictor with Elman neural networks for milling tool wear assessment with several statistic features selected from multi-sensor data including spindle current, force, vibration and acoustic emission.

The accuracy and generalization ability of the above-mentioned conventional feature-based machining learning models are highly affected by the quality of the hand-designed features [27]. However, feature selection is problem-dependent and somewhat subjective in practice. Different from feature-based machining learning methods, deep learning can achieve adaptive feature learning, which is helpful to improve the adaptability of prediction methods. Moreover, layer-by-layer feature learning in deep network is more likely to learn essential features hidden in the monitoring data and then to improve prediction accuracy [28]. Common deep learning methods include deep multi-layer perceptron (DMLP), deep autoencoder (DAE), convolutional neural network (CNN) and long short-term memory (LSTM) network. Serin et al. [29] used DMLP neural networks to predict surface roughness and specific energy consumption during 5-axis milling. Ou et al. [30] proposed an online sequential extreme learning machine for tool wear state recognition with a stacked denoising autoencoder (SDAE) put forward to extract abstract features. Cao et al. [31] proposed a 2-D CNN for milling tool wear monitoring, with the spectrum of the high signal-to-noise ratio vibration signals obtained from the derived wavelet frames as input features. Aghazadeh et al. [32] employed a CNN with a hybrid feature extraction method using wavelet time-frequency transformation and spectral subtraction algorithms for tool wear estimation. Mart'ınez-Arellano et al. [33] built a CNN model with time series imaging technique to transform the input raw signals. Sun et al. [34] designed an LSTM network to predict multiple flank wear values using raw signals of cutting force, vibration and acoustic emission. Zhao et al. [35] used convolutional bi-directional LSTM networks for tool condition monitoring in milling process.

From the above, most existing deep learning methods use raw signals without feature pre-selection and hand-design to maximize the merits of deep learning. In the actual machining process, cutting conditions are strongly time-varying, especially for the milling of parts with complex geometry, of which the allowance and cutting depth drastically change along the whole tool path [36]. In most cases, in the actual machining environment, only a few types of monitoring signals are possible to acquire, and the input features from the signals are probably highly correlated, leading to low prediction accuracy and poor generalization ability in time-varying cutting conditions. Besides, the model based on the raw signal as input has weak interpretability and it is difficult to analyze the source of error.

To solve this problem, a tool wear monitoring method for complex part milling based on deep learning is proposed in this paper. Compared with the existing approaches, the features are pre-selected with cutting force model and wavelet packet decomposition. The cutting depth, cutting forces, cutting force coefficients and the energy of the cutting vibration are selected as the input parameters and the output is the flank wear. Then, a deep autoencoder is employed to retract the highly correlated features from the pre-selected features, followed by a deep neural network to predict the wear value. The dataset for training and testing of the deep learning model is obtained with a carefully designed varying depth milling experiment. The testing results prove the effectiveness of the proposed method.

The remainder of this paper is organized as follows. The proposed monitoring method based on deep learning is introduced in Section 2. The experiment is conducted in Section 3. The application and results of the proposed method are shown and discussed in Section 4. The conclusions are summarized in Section 5.

2. Method

2.1. Overall Monitoring Method

Tool wear monitoring usually uses force, vibration, acoustic emission, current, power and temperature signals. In this study, to accord with the situation of few sensors in real machining environment, cutting force is selected as the single signal used for tool wear monitoring. In complex part milling, the cutting depth usually varies with the profile of the part. In this case, the cutting force and stability are time-varying, and it is difficult to design features for accurate tool wear estimation. To address this problem, a feature extractor based on pre-selection and deep learning is developed. The hand-designed features from the cutting force signal are further extracted with a deep autoencoder and then input to the deep multi-layer perceptron to estimate the tool wear value. The framework of the proposed tool wear monitoring method is illustrated in Figure 1.

2.2. Feature Pre-Selection

The features are pre-selected from the cutting force signals as the input of the deep learning, including nine features, i.e., the radial cutting depth, the magnitudes of two cutting force components in the tangential and axial direction, four cutting force coefficients and the two cutting vibration features in the tangential and axial direction.

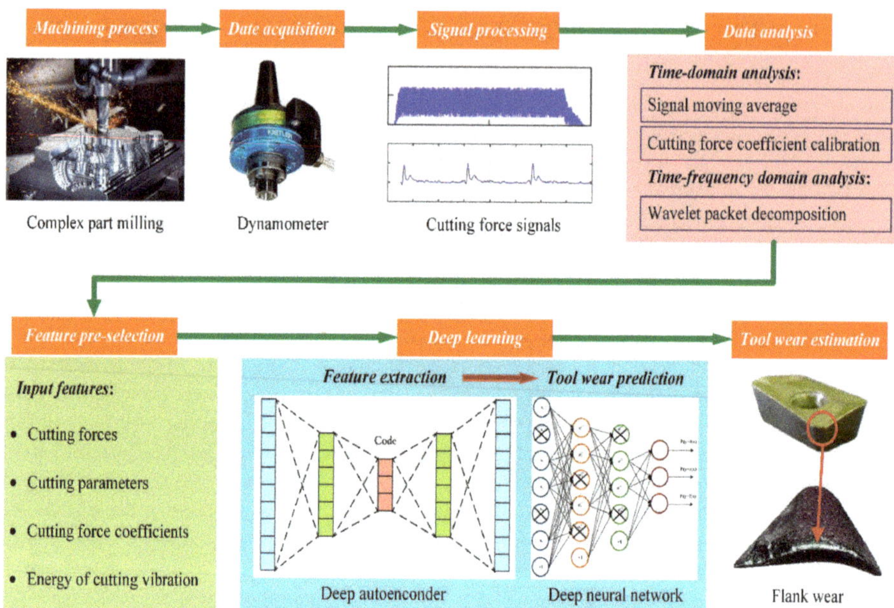

Figure 1. Framework of tool wear monitoring method for complex part milling based on deep learning.

2.2.1. Cutting Force Features

Cutting force is the most sensitive quantity reflecting the changes in tool wear. The magnitudes of cutting forces are selected as the features for tool wear monitoring. Cutting force coefficients can also be used as the cutting condition independent features for tool wear monitoring in complex part milling. Thus, the cutting force coefficients are also selected as the features.

The cutting forces in the milling process are illustrated in Figure 2. The cutting force model can be expressed as

$$\left\{ \begin{array}{l} F_t(\phi) = K_{tc}bh(\phi) + K_{te}b \\ F_r(\phi) = K_{rc}bh(\phi) + K_{re}b \\ F_a(\phi) = K_{ac}bh(\phi) + K_{ae}b \end{array} \right. \tag{1}$$

where F_t, F_r and F_a are the tangential, radial and axial cutting force, K_{tc}, K_{rc} and K_{ac} are the cutting force coefficients contributed by the shearing action in tangential, radial and axial directions, K_{te}, K_{re} and K_{ac} are the edge constants, ϕ is the instantaneous angle of immersion, b is the edge contact length, h is uncut chip thickness

$$h = f_z \sin \phi \tag{2}$$

where f_z is feedrate per tooth. According to Equations (1) and (2), the cutting force has a linear relationship with the uncut chip thickness h. When the spindle speed and feedrate are fixed, the uncut chip thickness h changes with the sine of the instantaneous angle of immersion ϕ. The linear function can be rewritten as

$$\left\{ \begin{array}{l} F_t/b = K_{tc}f_z \cdot \sin \phi + K_{te} \\ F_r/b = K_{rc}f_z \cdot \sin \phi + K_{re} \\ F_a/b = K_{ac}f_z \cdot \sin \phi + K_{ae} \end{array} \right. \tag{3}$$

As per Equation (3), the cutting force coefficients can be calibrated by linear regression of the cutting force value. Let the slope and intercept of the linear function after linear fitting be P_i and Q_i, respectively, the cutting force coefficients can be obtained as

$$\begin{cases} K_{ic} = P_i / f_z \\ \quad K_{ie} = Q_i \end{cases} \tag{4}$$

In complex part milling, the part is usually cut layer by layer in the axial direction. The tangential and radial force are varied significantly compared with the axial cutting force. Hence, in this study the average tangential and radial cutting forces F_t, F_r, and the tangential and radial cutting force coefficients K_{tc}, K_{rc}, K_{te}, K_{re} are selected as the input features.

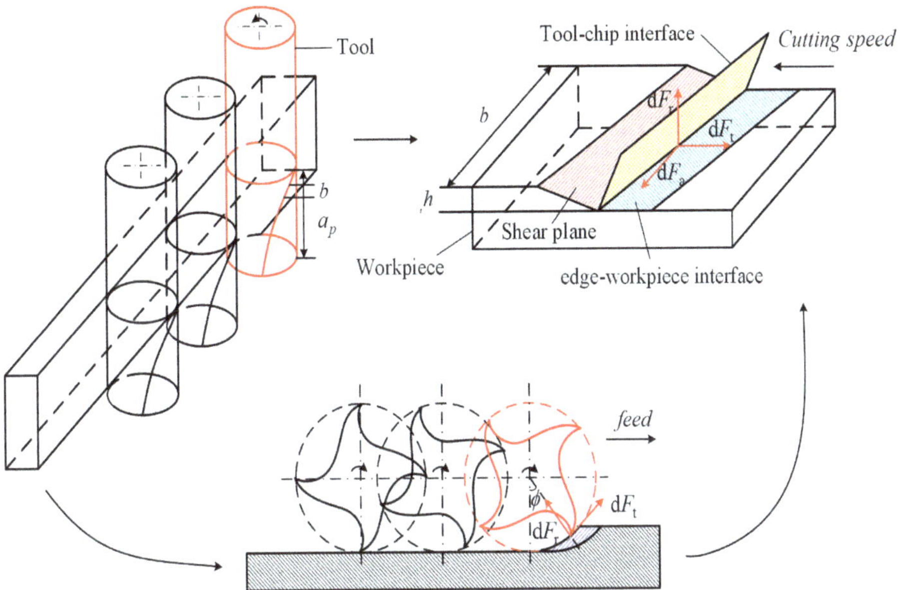

Figure 2. Interaction of tool and workpiece in milling process.

2.2.2. Cutting Vibration Features

Cutting vibration has high correlation with tool wear. In this study, the cutting vibration features are extracted from cutting force signals for tool wear monitoring. To this end, a three-layer wavelet packet decomposition is used to decompose the cutting force fluctuation components from the original cutting force signal, as shown in Figure 3.

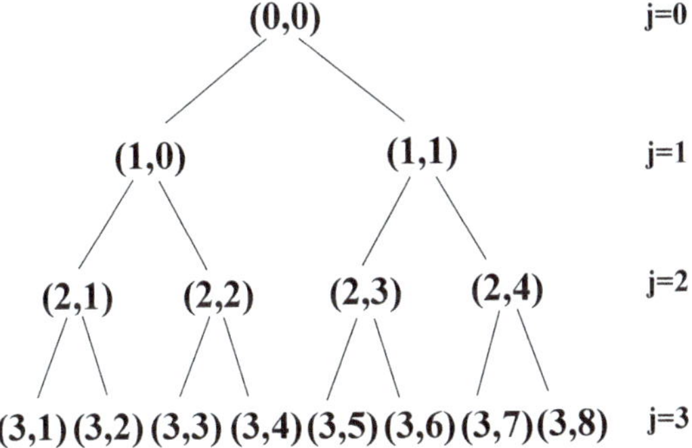

Figure 3. Structure of the three-layer wavelet packet decomposition for reconstruction of cutting force signals.

Wavelet packet decomposition (WPD) is a time-frequency analysis method. In WPD, the scale function of a standard orthogonalization $\psi(x)$ is used, and with two scale difference recursive Equations, the orthogonal wavelet packet is generated as

$$\begin{cases} w_{2n}(x) = \sqrt{2}\sum_{k=Z} h_k w_n(2x-k) \\ w_{2n+1}(x) = \sqrt{2}\sum_{k=Z} g_k w_n(2x-k) \end{cases} \tag{5}$$

where $w_0 = \psi(x)$, h_k, g_k are, respectively, a pair of conjugate quadrature filter coefficients derived from $\psi(x)$. For signal $s(t)$, the discrete orthogonal WPD is defined as the projection coefficient of $s(t)$ on the orthogonal wavelet packet base, which can be expressed as

$$P_s(n,j,k) = \langle s(t), w_{n,j,k}(t) \rangle = \int_{-\infty}^{+\infty} s(t)\left[2^{-\frac{j}{2w_n}}\left(2^{-j}t - k\right)\right]dt \tag{6}$$

where $\{P_s(n,j,k)\}$ is the WPD coefficient sequence of $s(t)$ on the orthogonal wavelet packet space U_j^n. By setting a group of different conjugate quadrature filter coefficients of $\{h_k\}$ and $\{g_k\}$, the wavelet packet transform coefficients can be expressed as

$$\begin{cases} P_s(2n,j,k) = \sum_{k=Z} h_{l-2k} P_s(n,j-1,l) \\ P_s(2n+1,j,k) = \sum_{k=Z} g_{l-2k} P_s(n,j-1,l) \end{cases} \tag{7}$$

Then, the energy distribution of the reconstructed signal in the time-frequency domain is expressed as

$$E(j,n) = \sum_{k=Z} [P_s(n,j,k)]^2 \tag{8}$$

In this study, the reconstructed force component with WPD caused by vibration is adopted as the input feature for tool wear monitoring.

2.3. Deep Learning for Tool Wear Monitoring

2.3.1. Structure of the Deep Learning Network

The deep learning network for tool wear monitoring in variable cutting depth milling is shown in Figure 4. It consists of two parts, a deep autoencoder and a deep multi-layer perceptron. The deep autoencoder is used to reduce the dimension of the input vectors (the pre-selected features) and learn the highly correlated features. Then, the following deep multi-layer perceptron is used to learn the effect of the extracted features on the tool wear and predict the tool wear value.

2.3.2. Deep Autoencoder

The pre-selected features are all obtained from cutting force signals. They are highly coupled and superfluous for tool wear prediction. This affects the accuracy and generalization ability of the deep learning model. In this study, the deep autoencoder (DAE) is used to learn the low-dimensional abstract features from the original features, which have a high correlation with the tool wear. The structure of the DAE is illustrated in Figure 5. Deep autoencoder is an unsupervised deep learning network for data dimensional reduction and feature extraction. It has a symmetric structure of multiple layers with the same input and output data. It consists of an encoder and a decoder. The transformation from the input layer to the middle hidden layer is called encoding. This is a dimensional reduction process, which transforms the high-dimensional original data into the low-dimensional space. The transformation from the middle hidden layer to the output layer is called decoding. This is the reconstruction process, which is the inverse of the encoder that reconstructs the encoded vector back to the original input data. In the encoding process, the obtained middle hidden layer reflects the essential features of the high-dimensional input data and is the core of the DAE. In this study, a six-layer DAE is developed with three layers for both the encoder and decoder.

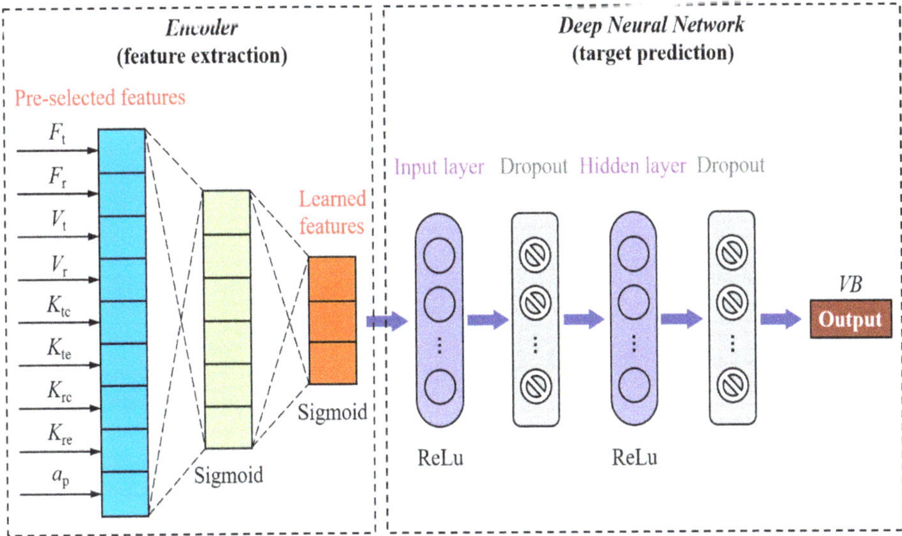

Figure 4. Architecture of deep learning network for tool wear prediction combining autoencoder and multi-layer perceptron.

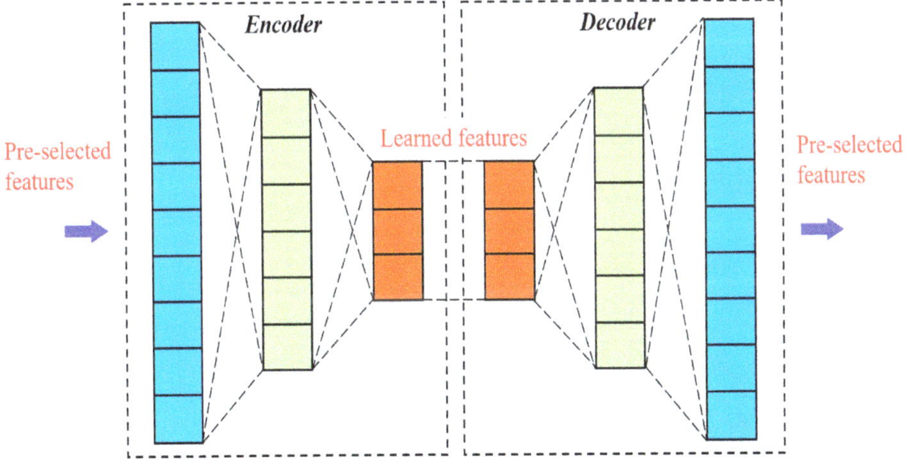

Figure 5. Symmetrical structure of deep autoencoder consisting of three-layer encoder and three-layer decoder.

The encoding process can be expressed as

$$h = f_\theta(x) = s_f(Wx + b) \tag{9}$$

The decoding process can be expressed as

$$\hat{x} = g_{\theta'}(h) = s_g\left(W' h + b'\right) \tag{10}$$

where $x = [x_1, x_2, \cdots, x_n]^T$ is the input vector, h and $\hat{x} = [\hat{x}_1, \hat{x}_2, \cdots, \hat{x}_n]^T$ are the encoded and decoded vectors, f_θ and $g_{\theta'}$ are the encoding and decoding functions, W is the weight matrix from the input layer to the hidden layer, W' is the weight matrix from the hidden layer to the input layer, b is the bias vector of the hidden layer, b' is the bias vector of the output layer, θ is the set of parameters in the encoding function, θ' is the set of parameters in the decoding function, s_f and s_g are the activation functions of the encoder and decoder, which are both selected as the sigmoid function.

In the training process, to measure the error between the input x and the reconstructed input x', the squared Euclidean distance is used as the loss function for this regression problem, which can be expressed as

$$Loss\left(W, b, W', b'\right) = \sum_{i=1}^{n} \|x_i - \hat{x}_i\|^2 \tag{11}$$

After the DAE has been trained, the encoder (the first three layers) is employed as a feature extractor in front of the deep learning model.

2.3.3. Deep Multi-Layer Perceptron

The deep multi-layer perceptron is used to predict tool wear with the input of the extracted features from the DAE. The deep multi-layer perceptron consists of an input layer, a hidden layer and an output layer. The input layer and hidden layer are followed by a dropout layer. The number of neurons in each layer is tuned in the training process. The Rectified Linear Unit (ReLU) is used as the activation function, which can be expressed as

$$f(x) = \max(0, x) \tag{12}$$

2.3.4. Loss Function

The tool wear prediction is a regression problem. The target of network training is to minimize the error between the predicted and measured wear values. In this study, the mean squared error (*MSE*) is used to evaluate the prediction error, which can be expressed as

$$MSE = \frac{1}{n}\sum_{i=1}^{n} \|y_i - \hat{y}_i\|^2 \tag{13}$$

where $y = [y_1, y_2, \cdots, y_n]^T$ is the measured output vector which refers to the tool wear value, $\hat{y} = [\hat{y}_1, \hat{y}_2, \cdots, \hat{y}_n]^T$ is the predicted output vector, and n is the number of samples.

2.3.5. Regularization

Regularization is an effective technique used to enhance the generalization ability of deep neural network. The L1 and L2 regularization are generally used regularization methods. This involves adding the L1 norm or the L2 norm of the weight to the loss function in order to constrain the weight of the neural network. In this way, the networks with large weight values are abandoned in training process. In this study, to accord with the MSE loss value, L2 regularization is adopted. Then, the final loss function of the deep neural network can be expressed as

$$Loss = \frac{1}{n}\sum_{i=1}^{n} \|y_i - \hat{y}_i\|^2 + \frac{\lambda}{2}\|w\|^2 \tag{14}$$

where $\|w\|$ is the L2 norm of the weight vector w, and λ is the regularization parameter. The regularization parameter is tuned in the training process.

Another regularization approach is Dropout. In the Dropout method, the neurons in the network are deleted randomly in the training process, as illustrated in Figure 6. This trick can significantly reduce the interaction between the feature detectors (the hidden layer nodes). Detector interaction means that, in a deep network, some detectors only show their effect depending on other detectors. In this study, the dropout layers are inserted behind the input and hidden layers to diminish overfitting in training process. The dropout rate is set as 0.2, which means each neuron has a 20% probability of being deleted in the training process.

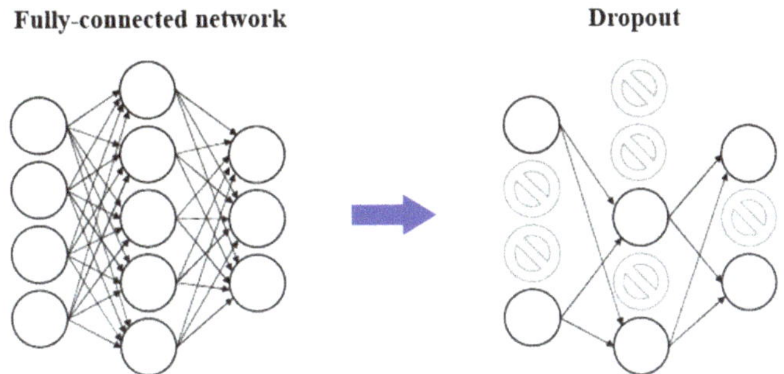

Figure 6. Schematic representation of dropout trick in a fully connected network.

3. Experiment

To acquire the training and testing dataset for a deep learning model, the milling experiment is performed. The experimental setup is shown in Figure 7. The experiment is carried out on a YHVT850Z three-axis machining center. The work material is GH4169. The up milling is used with a 12 mm diameter indexable carbide insert tool APMT113PDER-H2. To control random factors in the machining process, the milling process is performed without a coolant. The cutting force is measured with a Kistler 9123C rotating dynamometer. The flank wear of the cutting tool was measured with Alicona InfiniteFocus optical measurement device.

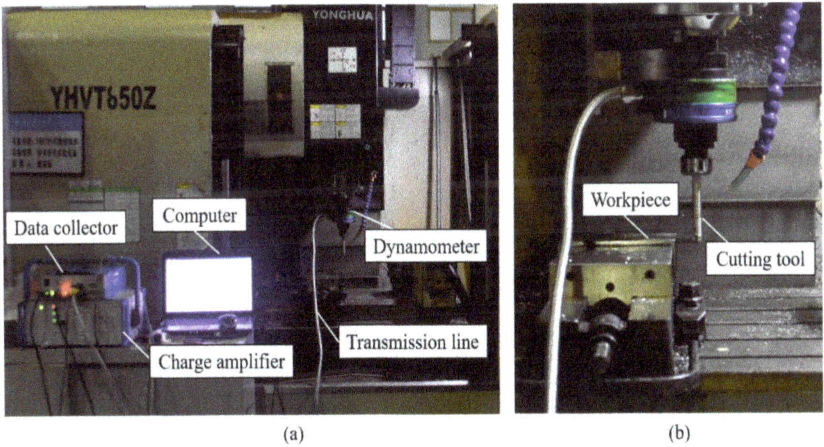

Figure 7. Experimental setup: (**a**) cutting force online measurement system, (**b**) layout of cutting tool and workpiece.

To simulate the actual milling process with time-varying cutting depth, the milling process is designed with variable radial cutting depth. The tool path is illustrated in Figure 8. A total of 12 cutting passes are performed. For each pass, the radial cutting depth is from 0.5 to 1.5 mm. With the increase of cutting length, tool wear is gradually increased. Each cutting pass is divided into 10 segments.

The cutting process was paused after each segment and the tool wear measurement was performed offline after each segment was finished. The average cutting forces in each segment are recorded as the input features of tool wear monitoring. The cutting parameters are listed in Table 1.

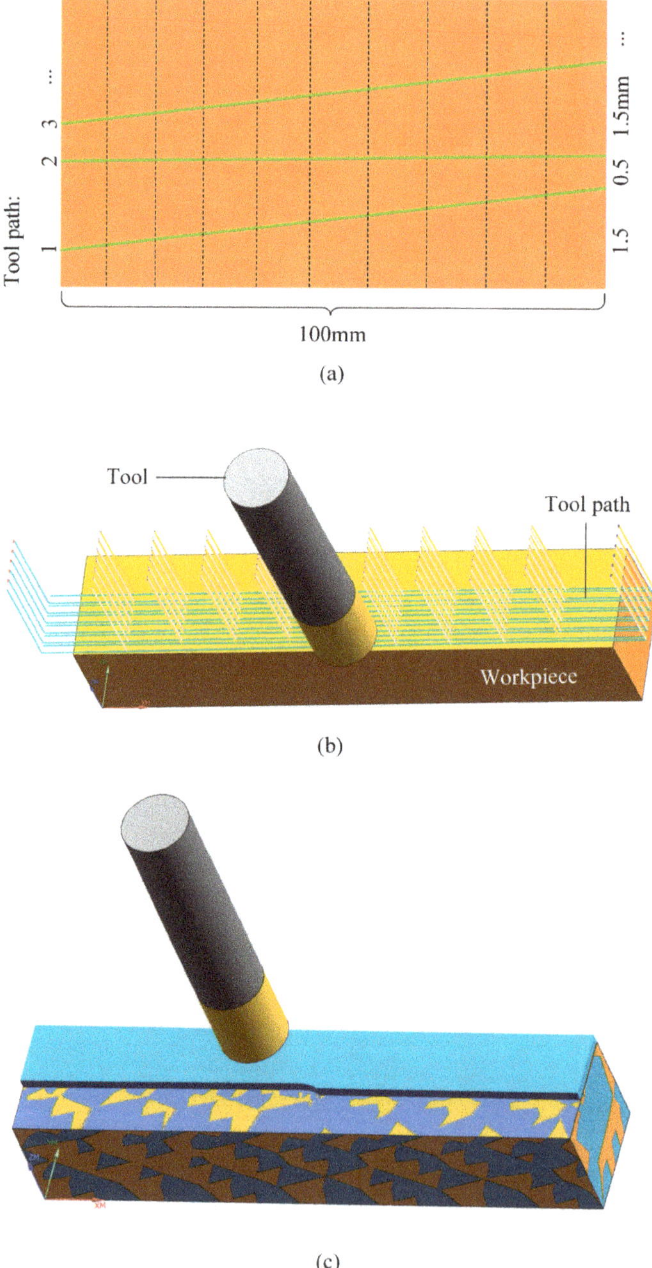

Figure 8. Tool path of varying depth milling: (**a**) tool path design, (**b**) tool path, (**c**) cutting process simulation.

Table 1. Cutting parameters in milling experiment.

Cutting Parameters	Setting Values
Spindle speed	1200 rev/min
Feedrate	0.05 mm/rev
Axial cutting depth	2 mm
Radial cutting depth	0.5~1.5 mm

4. Results and Discussion

4.1. Features Extracted from Cutting Force Signals

The raw cutting force signal in a single segment is shown in Figure 9. This shows that the magnitude of the force is increased with the cutting depth. The average cutting force and the cutting force coefficients are selected as the input cutting force features of the deep learning model for tool wear monitoring. The average value of the cutting force in each segment are calculated and used as the input features. Cutting force coefficients are calibrated by applying linear regression as per Equations (3) and (4).

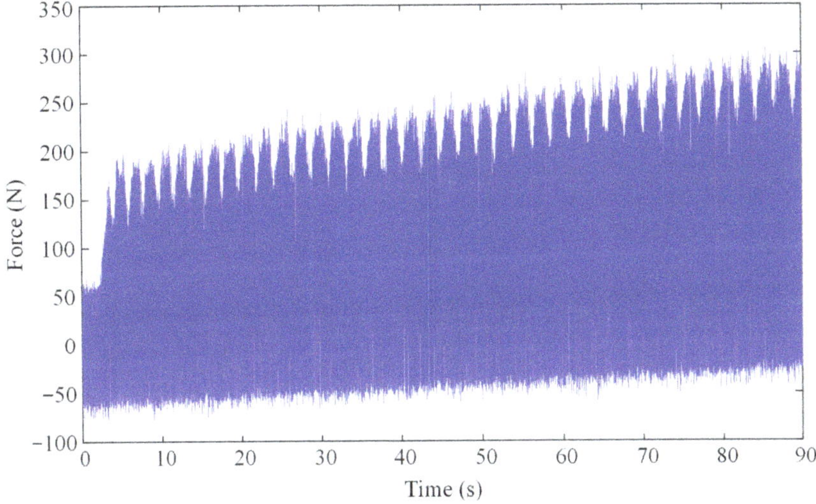

Figure 9. Raw signal of cutting force in a single segment.

By reconstructing the cutting force signal with the WPD, the cutting force components caused by cutting vibration are obtained. The features related to tool wear are extracted from these two kinds of signals. With the three-layer WPD, the cutting force signals in the 0~1.5 KHZ are decomposed into eight frequency bands. By using 8 db wavelet packet decomposition, the width of each frequency band is 150 Hz. Taking the cutting force signal in three cycles as an example, the reconstructed signals in each band after WPD are shown in Figure 10.

From Figure 10, the reconstructed signals in different frequency band can be analyzed intuitively. According to the cutting force model Equation (1), the signal in the 0~150 Hz frequency band is related to the static cutting force component; the other signal in higher frequency band can be regarded as the dynamic component caused by cutting vibration and noise. From the results of WPD, the tangential force signal in 150~300 Hz is related to cutting vibration. The reconstructed force is denoted as V_t. The radial force signal in 450~600 Hz is related to cutting vibration. The reconstructed force is denoted as V_r. These two parameters are selected as the input cutting vibration features of the deep learning model for tool wear monitoring.

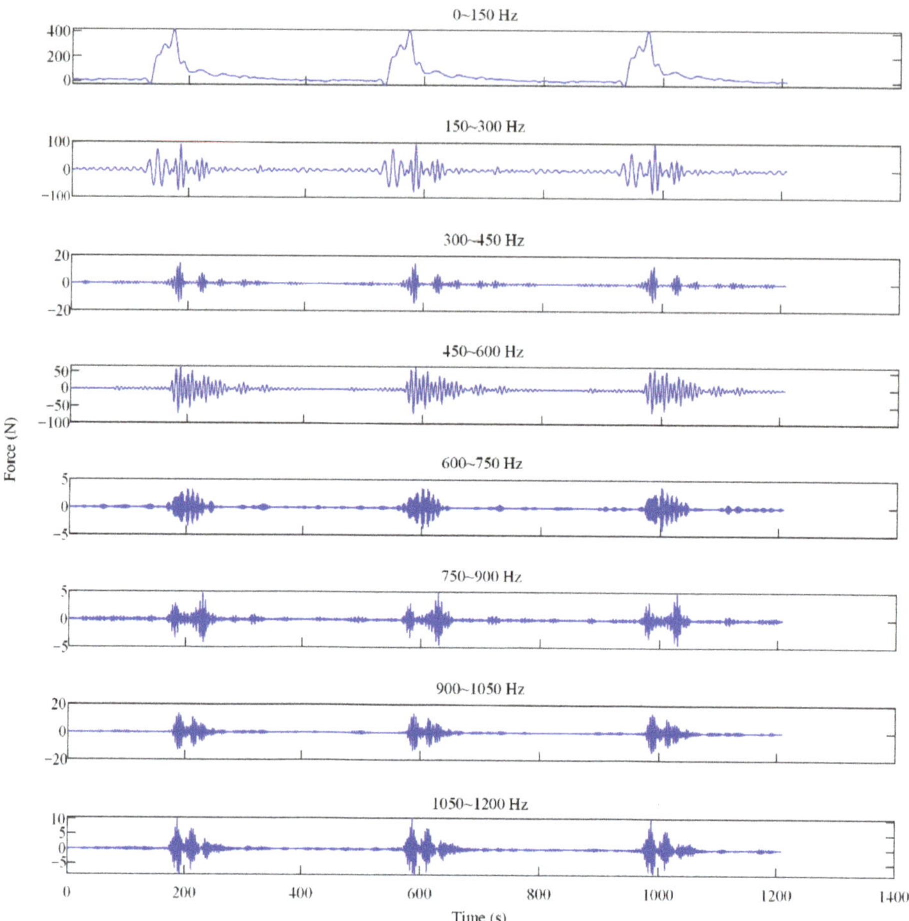

Figure 10. Reconstructed cutting force signals in each frequency band with wavelet packet decomposition.

4.2. Training and Validation of Deep Neural Network

After the feature pre-selection and measurement shown above, the dataset for deep learning is shown in Appendix A. There are a total of 120 samples (acquired from 10 segments in 12 cutting passes). The Keras deep learning library is employed with Tensorflow as the back-end to implement the proposed deep learning networks. The dataset is randomly split into 80% training set (96 samples) and 20% testing set (24 samples). Before training, all the input parameters are normalized. Then, the grid search technique is employed to select the optimal hyper-parameters, including epochs, neuron numbers, optimizer, learning rate and regularization parameter. The obtained optimal hyper-parameters are given in Table 2.

With the proposed deep learning model in Section 2, with the above optimal hyper-parameters, the training process is carried out. In the training process, four-fold validation is used, with 24 samples in each fold. The mean value of the mean absolute error in all the fold is recorded as the training error.

The training curve is shown in Figure 11. The final mean absolute error after 1000 epochs is 34.93 μm, and the corresponding R^2 value is 0.9826. This proves the accuracy of the trained deep learning model.

Table 2. Optimal hyper-parameters of deep neural network obtained by grid search.

Hyper-Parameters	Setting Values
Epochs	1000
Neuron number of the input layer	50
Neuron number of the hidden layer	10
Optimizer	RMSprop
Learning rate	0.001
Regularization parameter	0.001

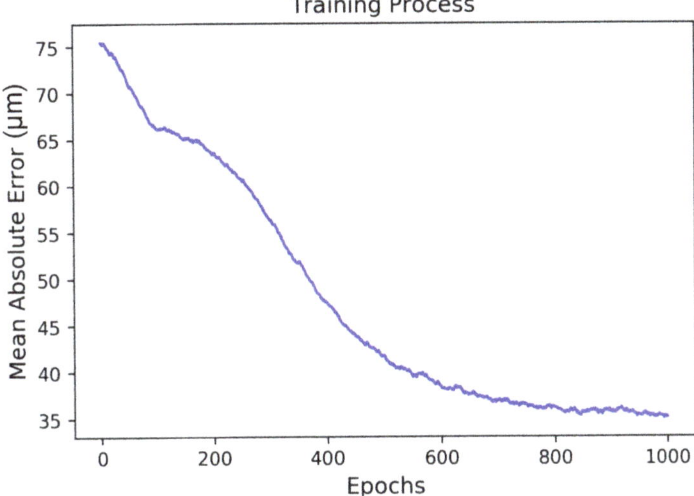

Figure 11. Mean absolute error of proposed deep learning model in training process.

The mean absolute error with the trained network in the four folds of the validation set is shown in Figure 12. The prediction error of tool wear value is less than 37 μm, and it is enough to monitor the tool wear state in the real machining process. This proves the effectiveness of the trained deep learning model.

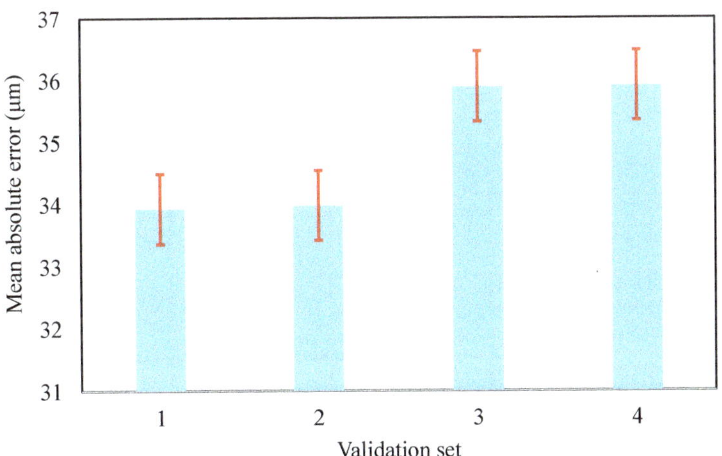

Figure 12. Mean absolute error in four-fold validation set.

4.3. Testing of Deep Neural Network

The trained deep learning model is used for tool wear prediction on the testing set. The average error in percentage is used to evaluate the performance of the trained model, which is calculated as

$$error = \frac{1}{m}\sum_{i=1}^{m} \frac{\left|VB_i^* - VB_i\right|}{VB_i} \tag{15}$$

where VB_i^* is the predicted tool wear value, VB_i is the measured tool wear value, and m is the total number of the testing samples.

The testing result is shown in Figure 13. The average error of the 24 testing samples is 8.2%, as per Equation (15). In the testing samples, the maximum relative error occurs in the test NO.13, with 64.8 μm larger than the measured value. In this case, the new tool has just begun to cut, with the measured flank wear of 25.7 μm. Besides, the maximum absolute error occurs in the test NO.15, with 82.3 μm less than the measured value. In this case, the cutting edge is seriously worn and the measured flank wear reaches 402.9 μm, far beyond the ISO tool wear criterion of 300 μm. Except for the two extreme cases shown above, where the actual tool wear value is too small or too large, the trained deep learning model predicts well for all the testing samples. For the cases where the tool wear is too small or too large, the model has a larger prediction error because the training set obtained from the experiment covers few samples of the tool wear with such extreme values, and thus the rules learnt in the model under these situations are somewhat under-fitting. The proposed deep learning model can be more accurate with more varied training data through additional experiments. This will be investigated in future research work. In this study, the testing result shows that the proposed model works well in most cases and this proves the generalization ability of the trained deep learning model.

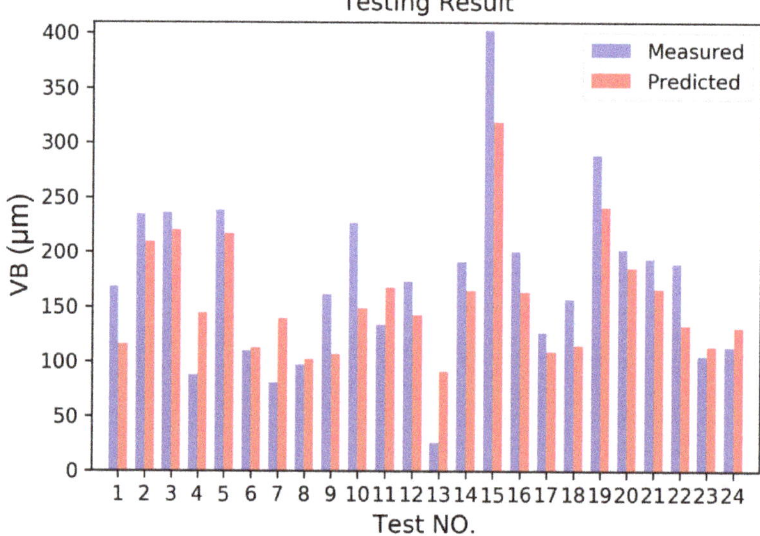

Figure 13. Testing result in 24 samples of the trained deep learning model.

Then, the radial basis function neural network (RBF NN) is implemented on the same dataset as the baseline to compare it with the performance of the proposed method. It turns out that the average error of RBF NN is 13.8%. Compared with the RBF NN, the proposed method shows higher accuracy in the tool wear prediction of varying cutting depth milling.

5. Conclusions

In this paper, a monitoring method based on deep learning is presented to give a solution to the feature extraction and accurate prediction of tool wear in complex part milling. A deep neural network is developed to estimate tool wear with the features pre-selected based on the cutting force model and wavelet packet decomposition, and further extracted using a deep autoencoder. The conclusions of this paper are summarized as: 1) the feature pre-selection based on cutting force model and wavelet packet decomposition and extraction coupled with a deep autoencoder is able to learn the highly correlated features for tool wear monitoring from single cutting force signals; 2) the proposed deep leaning model with the optimal settings has high accuracy for predicting tool wear; 3) the proposed deep learning method is proved to have higher accuracy than radial basis function neural network in varying cutting depth milling.

Author Contributions: Conceptualization, X.Z. and C.H.; methodology, X.Z. and C.H.; software, C.H.; validation, X.Z. and C.H.; formal analysis, M.L.; investigation, X.Z. and M.L.; resources, M.L. and D.Z.; data curation, M.L.; writing—original draft preparation, X.Z.; writing—review and editing, C.H. and M.L.; visualization, C.H.; supervision, D.Z.; project administration, D.Z.; funding acquisition, D.Z.. All authors have read and agreed to the published version of the manuscript.

Funding: This research was funded by National Natural Science Foundation of China, grant number 91860137, the Fundamental Research Funds for the Central Universities, grant number 31020200504002 and the Shaanxi Key Research and Development Project (Grant No. 2019KW-018).

Acknowledgments: Thanks are due to Tao Li for assistance with the experiments and data processing.

Conflicts of Interest: The authors declare no conflict of interest.

Appendix A

Table A1. Experimental data in varying cutting depth milling.

F_t (n)	F_r (n)	V_t (n)	V_r (n)	K_{tc} (MPa)	K_{te} (n/mm)	K_{rc} (MPa)	K_{re} (n/mm)	a_p (MPa)	VB (µm)
1459.1	298.1	1302.7	274.6	−1349.8	342	761.7	61	0.55	4.4381
1602.3	223.4	1413.4	330.1	−1238.5	176	649	73	0.65	13.7772
1425.2	269	1727.5	308.6	−973.21	237	606.69	29	0.75	25.6596
1683.9	209.7	1620.1	321.9	−1057.7	433	681.5	2	0.85	38.4103
1744.1	225.8	1629.4	301.7	−1054.1	294	687.3	80	0.95	50.89
1772.3	207.7	1660.3	232.4	−880.45	−156	730.55	−69	1.05	62.374
1672.8	177.4	1909.3	237.8	−889.18	−6	603.24	−38	1.15	72.4515
2003.3	168	1756.2	279.9	−878.06	250	642.87	13	1.25	80.9435
2096.7	380.3	1840.3	200.1	−844	357	618	−40	1.35	87.8359
2155.3	283	1947.2	289	−890.49	264	571.82	43	1.45	93.2272
1466.6	264.8	1469.2	283.9	−1220.2	345	1055.7	75	0.55	97.2865
1551.2	302.1	1552.2	338.8	−1350.9	260	823.4	150	0.65	100.2214
1634.5	316.1	1584.1	321.7	−1128.1	289	760.1	125	0.75	102.2543
1735.4	196.5	1601.5	326.3	−1138.4	447	742.2	54	0.85	103.6044
1719.9	243.2	1692.4	311	−1068.6	334	706.3	152	0.95	104.4762
1743.4	228.6	1740.2	235.3	−936.78	230	774.3	−31	1.05	105.0513
1831.6	184.2	1832.2	243.7	−952.43	−118	610.3	−46	1.15	105.4842
2014.8	226.2	1829.3	293.4	−951.5	288	647.7	27	1.25	105.9005
1981.3	245.4	1973.2	269	−919.93	372	630.93	73	1.35	106.3973
2208.2	262	2096.9	295.3	−979.94	243	641.1	78	1.45	107.0448
1397.1	287.6	1388.7	353.2	−1274.2	780	985.7	114	0.55	107.8888
1611.9	340.1	1616.2	380.5	−1314.4	570	856.4	168	0.65	108.9541
1670.2	331	1731.8	346	−1196.3	348	815.2	134	0.75	110.2482
1751.3	242	1788	321.5	−1264.7	511	757.6	155	0.85	111.7645
1865.5	262.9	1827	322.4	−1137.6	472	716.2	138	0.95	113.486

Table A1. *Cont.*

F_t (n)	F_r (n)	V_t (n)	V_r (n)	K_{tc} (MPa)	K_{te} (n/mm)	K_{rc} (MPa)	K_{re} (n/mm)	a_p (MPa)	VB (μm)
1867.6	267.5	1872.3	248.3	−1078.4	238	798.77	42	1.05	115.3889
1939.7	241.1	1978.7	276.1	−1104	204	646.73	35	1.15	117.4449
2055.8	337	2029.5	325.5	−1006.1	375	719.5	8	1.25	119.6239
2143.7	359	2306.2	390	−936.07	460	679.18	79	1.35	121.8962
2332.4	436.5	2429.1	464	−1000.4	336	670.88	129	1.45	124.2341
1478.4	380.2	1348.3	208.2	−1459.5	755	975.6	181	0.55	126.613
1774.5	378.7	1751.6	437.8	−1335	716	862.7	246	0.65	129.0121
1881.1	530.2	1918.5	544.6	−1362.7	562	827.4	199	0.75	131.415
2066.7	414.5	2093.3	519.6	−1346.9	754	801	187	0.85	133.8098
2004.8	413.3	1987.2	430.9	−1176.8	555	855.9	171	0.95	136.1888
2046.6	313.4	2003.2	263	−1142.4	342	809.5	88	1.05	138.5482
2094.5	289.4	2025.4	301.7	−1094.2	287	676.42	122	1.15	140.8876
2172.1	363	2302.2	384.7	−1031	378	730.4	25	1.25	143.2091
2262.7	440.5	2519.8	469.9	−904.63	374	791.85	177	1.35	145.5169
2497.5	558	2714.6	500.4	−1042.3	542	697.24	135	1.45	147.8161
1773.8	478.4	1560.1	462.6	−1118	−58	921.5	128	0.55	150.1124
1603.1	335.1	1714.2	384.8	−1273.9	305	622.4	240	0.65	152.411
1709.8	388.8	1758.7	419.4	−1190	333	572.7	250	0.75	154.7163
1806	331.6	1891.5	332.8	−1076.9	510	372.6	140	0.85	157.031
1842.8	293.9	1919.2	313.4	−913.35	147	351.8	190	0.95	159.3561
1902.8	342.9	2107.3	348.3	−946.33	268	155.59	289	1.05	161.6902
1952	368.9	2086.9	322.5	−990.2	246	263.8	202	1.15	164.0297
2018	260	2131.5	255.3	−829.85	−137	219.1	264	1.25	166.3684
2026.2	280.2	2192.9	315.1	−737.38	−249	297.1	246	1.35	168.698
2084.3	279.7	2075.5	206.4	−839.1	153	242.72	257	1.45	171.0079
1893.9	324.5	1768	475.8	−1134	63	1081.3	193	0.55	173.2859
1828.2	328.7	2010.7	404.2	−1304.5	407	780.5	244	0.65	175.5185
1889.2	397.5	2117.9	466.6	−1169.6	340	661	371	0.75	177.6913
2027.6	370.6	2159.4	410.8	−1186.2	553	352.2	257	0.85	179.7893
2007.6	325.6	2204.9	316.7	−1128.9	351	427.99	202	0.95	181.7983
2050.7	349.2	2272	396.9	−887.75	−26	406.65	302	1.05	183.7045
2142.9	196.3	2394.2	283.9	−1012.2	314	441.4	311	1.15	185.4958
2159.9	280.9	2494.5	313.3	−969.15	−35	348.69	369	1.25	187.1618
2188.6	312.7	2569.1	386.4	−888.6	55	396.01	394	1.35	188.6949
2305.8	291.3	2684	289.1	−911.61	391	310.12	366	1.45	190.09
1939	362.9	1843.1	481.2	−1673.4	681	1222.5	230	0.55	191.3455
1873.3	402.8	2098.1	460.2	−1272.2	624	991.1	334	0.65	192.4632
2000.5	472.5	2290.4	527.1	−1246.8	590	980.6	375	0.75	193.4483
1984.4	468	2322.7	399.9	−1197.3	411	1046.1	218	0.85	194.3099
2158.9	323.7	2368.6	338.3	−1120.1	490	938.9	219	0.95	195.0602
2153.3	369.4	2545.5	437.2	−1151.5	888	982	373	1.05	195.7152
2236.5	436.4	2458.5	372	−1048.5	360	913.1	315	1.15	196.2933
2286.6	276.3	2642.1	343.7	−1006.6	87	946.3	420	1.25	196.8157
2315.7	400.3	2752.6	421.1	−891.5	244	921.83	459	1.35	197.3056
2242.5	403	2722.2	403	−993.13	516	959.55	383	1.45	197.7874
1908	572.1	2238.1	523.5	−1654.1	818	1308.1	407	0.55	198.2861
1885.1	444.7	2268.6	595.1	−1389	798	1277.8	422	0.65	198.8267
1991.9	517.5	2353.8	657.6	−1379.1	640	1176.9	457	0.75	199.4334
2013.9	571.2	2434	469.2	−1367.8	912	1147.2	362	0.85	200.1284
2100.7	415.4	2582.5	537.4	−1315.1	1079	1099.8	347	0.95	200.9318
2075.2	379.4	2603.8	430.7	−1220.3	957	1136.4	395	1.05	201.8606
2196.2	460.6	2715.2	526.7	−1147.7	561	1083.2	433	1.15	202.9279
2033.4	388.4	2952.1	532.9	−1025.5	244	1056.3	484	1.25	204.1425
2242.2	488.3	2869.1	433.3	−1201.8	882	1089.5	489	1.35	205.5085
2332.3	468.7	2958.7	503.8	−916.35	187	994.5	480	1.45	207.0249

Table A1. *Cont.*

F_t (n)	F_r (n)	V_t (n)	V_r (n)	K_{tc} (MPa)	K_{te} (n/mm)	K_{rc} (MPa)	K_{re} (n/mm)	a_p (MPa)	VB (μm)
1867.7	333	2448.9	419.6	−1150.6	−144	1602	−83	0.55	208.6857
2062	330.7	2494.8	436.2	−1425.8	428	996.3	151	0.65	210.4795
2102	372	2609.9	452.3	−1264.1	278	1278.5	−399	0.75	212.3903
2208.2	401.5	2679.6	426.9	−1271.1	257	1412.5	21	0.85	214.3974
2234.8	424.5	2852.4	361.2	−947.3	442	1307.5	−251	0.95	216.4763
2310.1	281.8	2890.6	367.3	−1169.4	468	1250.1	−352	1.05	218.5993
2315.1	281.9	3005.4	354.5	−1100	481	1113.5	−34	1.15	220.7367
2477.8	299.7	3033.9	380	−1138.1	291	1164.8	−391	1.25	222.8577
2509.5	374.8	3144.9	446.4	−1068.8	120	1064.4	271	1.35	224.932
2315.8	371.7	3273.9	472	−844.1	100	1029.1	205	1.45	226.9309
2097.7	462.9	2633.9	478.2	−1524.8	161	1859	−515	0.55	228.8288
2144.6	376.8	2726.3	441.3	−1481.9	595	1566.8	225	0.65	230.605
2227	457.6	2820.1	521	−1393.5	223	1595.1	−268	0.75	232.245
2363.9	578.2	3011.5	432.5	−1355.6	276	1588.5	142	0.85	233.7418
2363.6	414.3	3061.6	425.4	−1059.3	525	1305.2	362	0.95	235.0976
2343.8	330.8	3220.2	375.3	−1182.6	504	1332.5	−11	1.05	236.3249
2468.6	370	3196.8	391	−1075.4	610	1328.8	146	1.15	237.447
2520.5	326.5	3389.9	444	−1259.4	435	1301.3	−242	1.25	238.4994
2500.6	291.2	3406.1	365	−1116.1	260	1284.5	197	1.35	239.5294
2536.6	451.7	3536.2	556.4	−1113.9	288	1257.9	−382	1.45	240.5966
2217.6	422	2809.5	477.3	−1355.7	894	1870.8	−191	0.55	241.7719
2315.8	482.7	2984.9	521.1	−1658.2	631	1895.2	475	0.65	243.1369
2300.7	531.4	3048	529.6	−1553	782	1589.9	297	0.75	244.7824
2418.9	627.2	3114.5	549.7	−1408.9	565	1681.8	178	0.85	246.8062
2407.4	689.2	3226.6	441.9	−1100.1	659	1605.6	573	0.95	249.3112
2575.1	418.9	3357.7	398.5	−1216.2	595	1512	345	1.05	252.4023
2469.1	474.1	3659.3	526.2	−1113.2	728	1360.8	657	1.15	256.1839
2551	343.3	3746.9	484.2	−1216.2	486	1354.6	624	1.25	260.7562
2627.1	400.8	3868	482.8	−1167.4	430	1358.7	438	1.35	266.2127
2598.8	349.2	3856.5	531.2	−1145.6	601	1212.1	553	1.45	272.6374
2221.1	598.5	2204.8	691.2	−1885.5	945	2161.7	−622	0.55	280.1025
2267.4	500.7	3340.9	617.1	−1592.5	1114	1972.7	626	0.65	288.668
2386.6	535.2	3390.2	570.9	−1646.5	1060	1646.8	414	0.75	298.3815
2597.7	796.9	3717.5	659.3	−1444.7	877	1691.9	527	0.85	309.2817
2563.2	707.3	3919.5	583.2	−1197.9	1233	1688.8	893	0.95	321.4035
2662.1	696.1	4051	621.3	−1374.5	678	1574.9	485	1.05	334.7875
2711.7	689.5	4156.5	513.8	−1266.3	954	1514.1	943	1.15	349.4937
2742.9	543.2	4333.8	640.3	−1348.7	526	1563.1	875	1.25	365.6212
2914.2	527.1	4622.7	492.5	−1448.2	518	1359.9	785	1.35	383.3351
2836.4	527.7	4773.4	674.9	−1157.2	665	1255.5	634	1.45	402.9015
1459.1	298.1	1302.7	274.6	−1349.8	342	761.7	61	0.55	4.4381
1602.3	223.4	1413.4	330.1	−1238.5	176	649	73	0.65	13.7772
1425.2	269	1727.5	308.6	−973.21	237	606.69	29	0.75	25.6596
1683.9	209.7	1620.1	321.9	−1057.7	433	681.5	2	0.85	38.4103
1744.1	225.8	1629.4	301.7	−1054.1	294	687.3	80	0.95	50.89
1772.3	207.7	1660.3	232.4	−880.45	−156	730.55	−69	1.05	62.374
1672.8	177.4	1909.3	237.8	−889.18	−6	603.24	−38	1.15	72.4515
2003.3	168	1756.2	279.9	−878.06	250	642.87	13	1.25	80.9435
2096.7	380.3	1840.3	200.1	−844	357	618	−40	1.35	87.8359
2155.3	283	1947.2	289	−890.49	264	571.82	43	1.45	93.2272
1466.6	264.8	1469.2	283.9	−1220.2	345	1055.7	75	0.55	97.2865
1551.2	302.1	1552.2	338.8	−1350.9	260	823.4	150	0.65	100.2214
1634.5	316.1	1584.1	321.7	−1128.1	289	760.1	125	0.75	102.2543
1735.4	196.5	1601.5	326.3	−1138.4	447	742.2	54	0.85	103.6044
1719.9	243.2	1692.4	311	−1068.6	334	706.3	152	0.95	104.4762

Table A1. *Cont.*

F_t (n)	F_r (n)	V_t (n)	V_r (n)	K_{tc} (MPa)	K_{te} (n/mm)	K_{rc} (MPa)	K_{re} (n/mm)	a_p (MPa)	VB (μm)
1743.4	228.6	1740.2	235.3	−936.78	230	774.3	−31	1.05	105.0513
1831.6	184.2	1832.2	243.7	−952.43	−118	610.3	−46	1.15	105.4842
2014.8	226.2	1829.3	293.4	−951.5	288	647.7	27	1.25	105.9005
1981.3	245.4	1973.2	269	−919.93	372	630.93	73	1.35	106.3973
2208.2	262	2096.9	295.3	−979.94	243	641.1	78	1.45	107.0448
1397.1	287.6	1388.7	353.2	−1274.2	780	985.7	114	0.55	107.8888
1611.9	340.1	1616.2	380.5	−1314.4	570	856.4	168	0.65	108.9541
1670.2	331	1731.8	346	−1196.3	348	815.2	134	0.75	110.2482
1751.3	242	1788	321.5	−1264.7	511	757.6	155	0.85	111.7645
1865.5	262.9	1827	322.4	−1137.6	472	716.2	138	0.95	113.486
1867.6	267.5	1872.3	248.3	−1078.4	238	798.77	42	1.05	115.3889
1939.7	241.1	1978.7	276.1	−1104	204	646.73	35	1.15	117.4449
2055.8	337	2029.5	325.5	−1006.1	375	719.5	8	1.25	119.6239
2143.7	359	2306.2	390	−936.07	460	679.18	79	1.35	121.8962
2332.4	436.5	2429.1	464	−1000.4	336	670.88	129	1.45	124.2341
1478.4	380.2	1348.3	208.2	−1459.5	755	975.6	181	0.55	126.613
1774.5	378.7	1751.6	437.8	−1335	716	862.7	246	0.65	129.0121
1881.1	530.2	1918.5	544.6	−1362.7	562	827.4	199	0.75	131.415
2066.7	414.5	2093.3	519.6	−1346.9	754	801	187	0.85	133.8098
2004.8	413.3	1987.2	430.9	−1176.8	555	855.9	171	0.95	136.1888
2046.6	313.4	2003.2	263	−1142.4	342	809.5	88	1.05	138.5482
2094.5	289.4	2025.4	301.7	−1094.2	287	676.42	122	1.15	140.8876
2172.1	363	2302.2	384.7	−1031	378	730.4	25	1.25	143.2091
2262.7	440.5	2519.8	469.9	−904.63	374	791.85	177	1.35	145.5169
2497.5	558	2714.6	500.4	−1042.3	542	697.24	135	1.45	147.8161
1773.8	478.4	1560.1	462.6	−1118	−58	921.5	128	0.55	150.1124
1603.1	335.1	1714.2	384.8	−1273.9	305	622.4	240	0.65	152.411
1709.8	388.8	1758.7	419.4	−1190	333	572.7	250	0.75	154.7163
1806	331.6	1891.5	332.8	−1076.9	510	372.6	140	0.85	157.031
1842.8	293.9	1919.2	313.4	−913.35	147	351.8	190	0.95	159.3561
1902.8	342.9	2107.3	348.3	−946.33	268	155.59	289	1.05	161.6902
1952	368.9	2086.9	322.5	−990.2	246	263.8	202	1.15	164.0297
2018	260	2131.5	255.3	−829.85	−137	219.1	264	1.25	166.3684

References

1. Liang, S.Y.; Hecker, R.L.; Landers, R.G. Machining process monitoring and control: The state-of-the-art. *J. Manuf. Sci. Eng.* **2004**, *126*, 297. [CrossRef]
2. Rehorn, A.G.; Jiang, J.; Orban, P.E. State-of-the-art methods and results in tool condition monitoring: A review. *Int. J. Adv. Manuf. Technol.* **2005**, *26*, 693–710. [CrossRef]
3. Choudhury, S.K.; Rath, S. In-process tool wear estimation in milling using cutting force model. *J. Mater. Process. Technol.* **2000**, *99*, 113–119. [CrossRef]
4. Cui, Y. Tool Wear Monitoring for Milling by Tracking Cutting Force Model Coefficients. Ph.D. Thesis, Shandong University, Jinan, China, 1997.
5. Shao, H.; Wang, H.L.; Zhao, X.M. A cutting power model for tool wear monitoring in milling. *Int. J. Mach. Tools Manuf.* **2004**, *44*, 1503–1509. [CrossRef]
6. Hou, Y.; Zhang, D.; Wu, B.; Luo, M. Milling Force Modeling of Worn Tool and Tool Flank Wear Recognition in End Milling. *IEEE/ASME Trans. Mechatronics* **2015**, *20*, 1024–1035. [CrossRef]
7. Han, C.; Luo, M.; Zhang, D.; Wu, B. Mechanistic modelling of worn drill cutting forces with drill wear effect coefficients. *Procedia CIRP* **2019**, *82*, 2–7. [CrossRef]
8. Han, C.; Luo, M.; Zhang, D. Optimization of varying-parameter drilling for multi-hole parts using metaheuristic algorithm coupled with self-adaptive penalty method. *Appl. Soft Comput.* **2020**, *95*, 106489. [CrossRef]

9. Han, C.; Zhang, D.; Luo, M.; Wu, B. Chip evacuation force modelling for deep hole drilling with twist drills. *Int. J. Adv. Manuf. Technol.* **2018**, *98*, 3091–3103. [CrossRef]

10. Han, C.; Luo, M.; Zhang, D.; Wu, B. Iterative Learning Method for Drilling Depth Optimization in Peck Deep-Hole Drilling. *J. Manuf. Sci. Eng.* **2018**, *110*, 121009. [CrossRef]

11. Yu, J. Tool condition prognostics using logistic regression with penalization and manifold regularization. *Appl. Soft Comput.* **2018**, *64*, 454–467. [CrossRef]

12. Kilickap, E.; Yardimeden, A.; Çelik, Y.H. Mathematical Modelling and Optimization of Cutting Force, Tool Wear and Surface Roughness by Using Artificial Neural Network and Response Surface Methodology in Milling of Ti-6242S. *Appl. Sci.* **2017**, *7*, 1064. [CrossRef]

13. Patra, K.; Jha, A.K.; Szalay, T.; Ranjan, J.; Monostori, L. Artificial neural network based tool condition monitoring in micro mechanical peck drilling using thrust force signals. *Precis. Eng.* **2017**, *48*, 279–291. [CrossRef]

14. Karam, S.; Centobelli, P.; D'Addona, D.M.; Teti, R. Online Prediction of Cutting Tool Life in Turning via Cognitive Decision Making. *Procedia CIRP* **2016**, *41*, 927–932. [CrossRef]

15. Venkatarao, K.; Murthy, B.; Rao, N.M. Prediction of cutting tool wear, surface roughness and vibration of work piece in boring of AISI 316 steel with artificial neural network. *Measurement* **2014**, *51*, 63–70. [CrossRef]

16. D'Addona, D.M.; Ullah, A.M.M.S.; Matarazzo, D. Tool-wear prediction and pattern-recognition using artificial neural network and DNA-based computing. *J. Intell. Manuf.* **2017**, *28*, 1285–1301. [CrossRef]

17. Mikołajczyk, T.; Nowicki, K.; Bustillo, A.; Pimenov, D.Y. Predicting tool life in turning operations using neural networks and image processing. *Mech. Syst. Signal Process.* **2018**, *104*, 503–513. [CrossRef]

18. Madhusudana, C.K.; Kumar, H.; Narendranath, S. Face milling tool condition monitoring using sound signal. *Int. J. Syst. Assur. Eng. Manag.* **2017**, *8*, 1643–1653. [CrossRef]

19. Benkedjouh, T.; Medjaher, K.; Zerhouni, N.; Rechak, S. Health assessment and life prediction of cutting tools based on support vector regression. *J. Intell. Manuf.* **2015**, *26*, 213–223. [CrossRef]

20. Zhang, C.; Zhang, H. Modelling and prediction of tool wear using LS-SVM in milling operation. *Int. J. Comput. Integr. Manuf.* **2016**, *29*, 76–91. [CrossRef]

21. Yu, J.; Liang, S.; Tang, D.; Liu, H. A weighted hidden Markov model approach for continuous-state tool wear monitoring and tool life prediction. *Int. J. Adv. Manuf. Technol.* **2017**, *91*, 201–211. [CrossRef]

22. Zhu, K.; Liu, T. Online Tool Wear Monitoring Via Hidden Semi-Markov Model with Dependent Durations. *IEEE Trans. Ind. Inform.* **2018**, *14*, 69–78. [CrossRef]

23. Tobon-Mejia, D.A.; Medjaher, K.; Zerhouni, N. CNC machine tool's wear diagnostic and prognostic by using dynamic Bayesian networks. *Mech. Syst. Signal Process.* **2012**, *28*, 167–182. [CrossRef]

24. Kong, D.; Chen, Y.; Li, N. Gaussian process regression for tool wear prediction. *Mech. Syst. Signal Process.* **2018**, *104*, 556–574. [CrossRef]

25. Wu, D.; Jennings, C.; Terpenny, J.; Gao, R.X.; Kumara, S. A Comparative Study on Machine Learning Algorithms for Smart Manufacturing: Tool Wear Prediction Using Random Forests. *J. Manuf. Sci. Eng.* **2017**, *139*. [CrossRef]

26. Liu, Y.; Wang, F.; Lv, J.; Wang, X. A Novel Method for Tool Identification and Wear Condition Assessment Based on Multi-Sensor Data. *Appl. Sci.* **2020**, *10*, 2746. [CrossRef]

27. Serin, G.; Sener, B.; Ozbayoglu, A.M.; Unver, H.O. Review of tool condition monitoring in machining and opportunities for deep learning. *Int. J. Adv. Manuf. Technol.* **2020**, *109*, 953–974. [CrossRef]

28. Sun, C.; Ma, M.; Zhao, Z.; Tian, S.; Yan, R.; Chen, X. Deep Transfer Learning Based on Sparse Autoencoder for Remaining Useful Life Prediction of Tool in Manufacturing. *IEEE Trans. Ind. Inform.* **2019**, *15*, 2416–2425. [CrossRef]

29. Serin, G.; Gudelek, M.U.; Ozbayoglu, A.M.; Unver, H.O. Estimation of Parameters for the Free-Form Machining with Deep Neural Network. In Proceedings of the 2017 IEEE International Conference on Big Data (Big Data), Boston, MA, USA, 11–14 December 2017; IEEE: Piscataway, NJ, USA, 2017; pp. 2102–2111.

30. Ou, J.; Li, H.; Huang, G.; Yang, G. Intelligent Analysis of Tool Wear State Using Stacked Denoising Autoencoder with Online Sequential-Extreme Learning Machine. *Measurement* **2020**, 108153. [CrossRef]

31. Cao, X.; Chen, B.; Yao, B.; Zhuang, S. An Intelligent Milling Tool Wear Monitoring Methodology Based on Convolutional Neural Network with Derived Wavelet Frames Coefficient. *Appl. Sci.* **2019**, *9*, 3912. [CrossRef]

32. Aghazadeh, F.; Tahan, A.; Thomas, M. Tool condition monitoring using spectral subtraction and convolutional neural networks in milling process. *Int. J. Adv. Manuf. Technol.* **2018**, *98*, 3217–3227. [CrossRef]

33. Martínez-Arellano, G.; Terrazas, G.; Ratchev, S. Tool wear classification using time series imaging and deep learning. *Int. J. Adv. Manuf. Technol.* **2019**, *104*, 3647–3662. [CrossRef]
34. Sun, H.; Zhang, J.; Mo, R.; Zhang, X. In-process tool condition forecasting based on a deep learning method. *Robot. Cim. Int. Manuf.* **2020**, *64*, 101924. [CrossRef]
35. Zhao, R.; Yan, R.; Wang, J.; Mao, K. Learning to Monitor Machine Health with Convolutional Bi-Directional LSTM Networks. *Sensors* **2017**, *17*, 273. [CrossRef] [PubMed]
36. Han, C.; Zhang, D.; Wu, B.; Pu, K.; Luo, M. Localization of freeform surface workpiece with particle swarm optimization algorithm. In Proceedings of the 2014 International Conference on Innovative Design and Manufacturing (ICIDM), Montreal, QC, Canada, 13–15 August 2014; IEEE: Piscataway, NJ, USA, 2014; pp. 47–52.

MDPI
St. Alban-Anlage 66
4052 Basel
Switzerland
Tel. +41 61 683 77 34
Fax +41 61 302 89 18
www.mdpi.com

Applied Sciences Editorial Office
E-mail: applsci@mdpi.com
www.mdpi.com/journal/applsci